P9-DBZ-878

Case Studies in Medical Physiology

Robert S. Alexander, Ph.D.
Professor of Physiology, Albany Medical College of Union University,
Albany, New York

Little, Brown and Company   Boston

Library of Congress Catalog Card No. 77-70460

ISBN 0-316-03124-0

Printed in the United States of America

Preface

To distinguish normal physiology from pathophysiology is to fail to understand disease. Most of the functional disturbances a physician encounters at the bedside are manifestations of normal physiological mechanisms responding to the stresses of disease. These natural stresses are not fundamentally different from the artificial ones that the physiologist employs in the laboratory to analyze physiological function. A patient who is running a spiking fever, for example, provides an excellent opportunity for the study of how heat conservation and heat loss mechanisms regulate body temperature.

Unfortunately, the complexity of human disease and the priority that must be given to the well-being of the patient often obscure the nature of the physiological processes. The beginning student will therefore be wisely advised to confine his attention to such simplified models as squid axones, turtle hearts, and segments of guinea pig intestine until he has mastered the fundamental principles of physiological function. Yet achieving this mastery is a sterile accomplishment for the medical student unless these principles can be related to bedside medicine.

This book is designed to take the reader to the bedside and challenge him to identify the physiology encountered there. It should prove useful for the student who is just completing his basic mastery of the subject, as well as for the more advanced student or physician who wishes to review the physiology of disease in a context divorced from the practical problems of therapy and patient management. Knowledge of the physiology of the patient is essential for understanding of his disease, but it should be clearly recognized that physiology alone does not tell us how to practice medicine effectively.

The reader should be forewarned that these case histories have been severely edited to delete aspects that were not specifically relevant to the physiological problem. Because of these deletions, there is no pretense that the case descriptions in this book present acceptable diagnostic work-ups or appropriate therapeutic management of these clinical problems, matters that are remote from the author's competence.

Two departures from conventional units have been incorporated into the presentation of the data. Years of classroom frustration in teaching acid-base physiology have convinced me that a major source of difficulty in understanding this subject is the confusion inherent in the inverted log scale of pH. Acidity data are therefore expressed directly in units of nanomoles (nM) of hydrogen ion and can simply be manipulated by use of the linear form of the Henderson equation:

$PCO_2 = (H^+ \times HCO_3^-)/24$, in which the PCO_2 is expressed in mm Hg and the HCO_3^- is in units of millimoles. Since the normal plasma concentration of HCO_3^- is 24 mM, this system starts from the happy coincidence that both H^+ and PCO_2 have as their normal reference level the identical numerical value of 40. For readers familiar with the pH scale but unfamiliar with the H^+ scale, a conversion table is provided at the back of the book. Similarly, the widely used but potentially ambiguous "milliequivalent" unit has been replaced by the millimolar unit (mM). For readers not familiar with normal blood values, a table of these is also provided at the rear of the book.

The cases have been selected from a file of case histories that I have collected over a number of years because of their physiological interest. They have been grouped into eight sections to identify the system to which they relate. Questions are appended to each case history to encourage the reader to work through his own analysis of the case before accepting the crutch of the author's analysis.

I am deeply indebted to many clinical colleagues and upper-class medical students who have called these cases to my attention.

R. S. A.

Contents

Respiration

I

Police were called to a college fraternity house early one morning. They found the body of a student at the shallow end of a swimming pool in about 3 feet of water. Near the head of the student was the end of a length of garden hose, 1 inch in diameter and 6 feet long; the opposite end of the hose had been tied to a ladder at the edge of the pool several inches above the water line.

Questioning of the student's fraternity brothers revealed that during a party the previous evening some sort of a hide-and-seek game was initiated. Apparently the dead student had been a participant in this game and had not been seen since that time. Detailed reports were somewhat confused because appreciable amounts of alcohol had been consumed by all concerned. It was theorized that the student assumed he could hide at the bottom of the swimming pool by breathing through the length of garden hose.

Questions

What problems would be encountered in maintaining respiration under these conditions?

Would they have been of sufficient magnitude to account for the death of the student?

Analysis of Case 1

Two factors would have made it impossible for the student to maintain respiratory exchange under these conditions. A length of hose 6 feet long by 1 inch in diameter would contain a volume of approximately 1 liter, which would represent added dead space to the respiratory circuit. In the normal situation, an alveolar ventilation of about 350 ml per breath is achieved by a tidal volume of 500 ml, 150 ml of which is wasted in normal dead-space ventilation. To sustain the same alveolar ventilation with the dead space increased by 1 liter would require a threefold increase in tidal volume.

Complicating this requirement is an even more serious pressure problem. The musculature of the thoracic cage is designed as a bellows to move substantial air volumes at low pressures. In normal function these low pressures are sufficient to overcome elastic forces in the lungs of the order of 6 to 8 mm Hg and a small resistance to air flow that creates an additional pressure gradient of 1 to 2 mm Hg. The maximal inspiratory pressure the respiratory muscles are capable of developing against a closed airway is some ten times this amount, or about 90 mm Hg at the full expiratory position. This capability for developing inspiratory pressure decreases significantly as the chest expands. The 3 feet of water in which the student was immersed represents a hydrostatic pressure equivalent to almost one-tenth of an atmosphere, or 70 mm Hg. To expand the chest by inspiring gas at atmospheric pressure, the respiratory muscles would have to create this additional pressure to displace the water surrounding the chest. An additional complication was created by the effects of the student's supine posture and the external hydrostatic pressure on the soft tissues of the body, which would counteract the normal gravitational forces tending to pool blood in the extremities. This would shift a significant volume of blood into his chest to congest the pulmonary vascular bed and decrease the compliance of his lungs. The student would, therefore, have been incapable of achieving any significant inspiration under these circumstances, and most certainly not the large tidal volume required to compensate for the dead space of the hose.

A 28-year-old man was admitted to the hospital for evaluation of his complaint of chronic cough. His examination included the initial pulmonary function studies listed in the table below. On the morning when he was scheduled for discharge, he suddenly had a severe bout of coughing and became extremely short of breath and cyanotic. Physical and fluoroscopic examination established the diagnosis of pneumothorax, and the patient was returned to bed rest. For academic purposes, the pulmonary function studies were repeated that afternoon. Examination of the patient two weeks later showed physical findings and an x-ray of the chest that were indistinguishable from those taken at the time of his original admission.

Pulmonary function studies	Expected normal	Initial study	Repeat study
Respiration rate (breaths/min)	12	9	22
Tidal volume (ml)	600	760	240
Inspiratory reserve (ml)	3000	1880	1630
Expiratory reserve (ml)	1200	2330	610
Vital capacity (ml)	4800	4970	2480
Functional residual capacity (ml)	2400	3420	920
Forced expiratory volume in 1 sec (% VC)	85	40	31
Lung compliance (liters/cm H_2O)	0.20	0.19	0.08
Chest compliance (liters/cm H_2O)	0.20	0.17	0.19
Airway resistance (cm H_2O/ liter/sec)	1.0	3.2	4.2
Arterial saturation (%)	97	96	93

Questions
Explain the abnormalities in pulmonary function observed in the initial examination.

What accounts for the changes observed after the pneumothorax?

What happened to the pneumothorax during the subsequent two weeks?

Analysis of Case 2

The initial pulmonary studies of this patient can be explained as the manifestations of increased airway resistance, which could be compatible with chronic bronchitis. Vital capacity was a bit greater than the expected normal, a not uncommon finding in patients with moderate resistive airway disease, since their respiratory musculature has been well developed by the excess demand placed on it. The distribution of this vital capacity was significantly abnormal, however. The small inspiratory reserve and large expiratory reserve indicate that the tidal volume was riding up toward the full inspiratory position. Associated with this was a prolongation of the time required to exhale the vital capacity, with only 40% moved during the first second. All these elements reflect the fact that airway resistance impedes expiration more than inspiration. Resistance is minimal during inspiration, when the negative intrathoracic pressure dilates the bronchial tree, and becomes maximal during forced expiration, when positive intrathoracic pressures narrow the airway. The direct measurement of airway resistance confirmed this interpretation of the spirometric examination.

Patients with increased airway resistance learn that rapid expiratory efforts are futile because they lead to collapse of the smaller airways that lack cartilagenous support. It is, therefore, of interest that this patient exhibited a relatively slow respiratory rate with a correspondingly higher tidal volume than normal, which would reduce the tendency for airway collapse and require less work in breathing. The reduced chest compliance that was observed would relate to the patient's large functional residual capacity. The inflated chest position in which the patient is doing his tidal respiration represents a range of reduced distensibility of the chest wall as compared with that at normal lung volumes.

Hard coughing in the presence of obstructive bronchial disease makes a patient vulnerable to developing high pressures within a segment of the lung, which may tear the pleural lining. The pneumothorax this creates will produce relatively more positive intrathoracic pressures, which will further aggravate the airway narrowing and add to the already high airway resistance. In addition, the collapsed lung requires much greater pressures to inflate. Both the fall in the vital capacity and the drop in pulmonary compliance reflect this difficulty of inflating the collapsed lung, overinflating the opposite lung, or both.

Note that at the time of the acute episode the patient was described as being cyanotic, but by the time of the laboratory study the arterial saturation was almost normal. Blood flow to unventilated alveoli is sharply curtailed by a combination of neural, chemical, and humoral factors to shunt blood toward well-ventilated alveoli, thereby minimizing arterial desaturation.

The fate of a pneumothorax, or any other collection of gas trapped within body tissues, is explained by the equilibration of this gas with the surrounding tissue fluids. These fluids will have gas tensions of the same order as systemic capillary blood, which in turn should be the same as those in venous blood. At sea level, venous blood has a nitrogen tension of 573 mm Hg, an oxygen tension that averages 40 mm Hg, and a carbon dioxide tension of 46 mm Hg. Adding the water-vapor tension of 47 mm Hg yields a total gas tension of 706 mm Hg, some 54 mm Hg below the atmospheric pressure. This negative pressure, which develops in the trapped gas volume as it tends to equilibrate with the surrounding tissue fluid, is a very effective force to pull out a collapsed lung or, in other body tissues, to pull soft tissues into the gas cavity. Since the consequent reduction in the gas volume will increase its pressure, the total gas tension in the trapped gas will tend to rise above 706 mm Hg toward atmospheric tension. This rise will proportionately elevate the partial pressure of each of the contained gases to levels above the partial pressure of the same gases in the surrounding tissues. The trapped gas will, therefore, progressively diffuse into the tissues until the gas pocket disappears entirely. This "vacuum cleaner" characteristic of venous blood finds its explanation in the unique form of the hemoglobin dissociation curve. Blood oxygen tension falls some 60 mm Hg as oxygen is delivered to the tissues, while an essentially equivalent amount of carbon dioxide is picked up in the tissues with a rise of only 6 mm Hg in carbon dioxide tension in the venous blood. This large discrepancy in the changes in the tensions of the two respiratory gases creates the subatmospheric pressure in the venous blood.

3

A 24-year-old veteran contracted pulmonary tuberculosis shortly after entering the service and was treated for active disease for three years. X-ray examinations during this period demonstrated diffuse infiltrative lesions through all lung fields. His period of active disease was also complicated by repeated bouts of pleurisy. He had been free of evidence of active disease for the subsequent two years, but was seriously incapacitated by dyspnea on mild exertion. In current x-rays, there were multiple calcified scars throughout his left lung field, while the right side of his chest was virtually opaque to x-rays.

Analysis of the patient's pulmonary function in the laboratory gave the data shown in the table.

Pulmonary function studies	Expected normal	Observed
Respiratory rate (breaths/min)	12	19
Tidal volume (ml)	600	430
Inspiratory reserve (ml)	2200	400
Expiratory reserve (ml)	900	430
Vital capacity (ml)	3700	1260
Forced expiratory volume in 1 sec (% VC)	85	88
Maximum ventilatory volume (liters/min)	159	67
Arterial saturation — at rest (%)	95	89
Arterial saturation — exercise	97	88
Arterial saturation — breathing O_2	100	92

Bronchospirometry demonstrated that his right lung was responsible for 16% of his tidal volume, 18% of his resting oxygen consumption, and 9% of his vital capacity. An esophageal balloon was passed, which recorded a pressure of -4 cm H_2O at the end of expiration with atmospheric airway pressure. He was then instructed to inspire 500 ml of air from a spirometer. The expiratory valve in the spirometer circuit was closed, and he was told to relax his respiratory muscles completely. Esophageal pressure measured -7 cm H_2O and airway pressure rose to $+9$ cm H_2O when he relaxed his inspiratory muscles.

The patient was referred to surgery.

Questions

Analyze each of the measured variables of pulmonary function and explain its physiological implications.

What is the nature of the functional deficiency in this patient?

Analysis of Case 3

The clinical picture of this patient suggested that his right lung was encased in a fibrous scar, resulting from his repeated bouts of tuberculous pleurisy, and that this scar was restricting lung expansion and accounting for his exertional dyspnea. In view of the evidence of extensive scarring from old disease in the left lung, however, the surgeon wished to assess the functional status of this lung before committing himself to attempting resection of the fibrous scar that was restricting right lung expansion.

The deficiency in vital capacity and maximum ventilatory volume indicated severe mechanical restriction to ventilation. The low expiratory reserve volume, the high value (in percent of vital capacity) of his 1 sec forced expiratory volume, and the fact that his maximal ventilatory volume was proportionately greater than his vital capacity all indicate normal airways with low airway resistance. The deficiency in arterial oxygen saturation seems to reflect some shunting of blood past scarred areas of lung tissue. If it had been a sign of any generalized impairment in pulmonary diffusion capacity, we would have expected a much greater rise in saturation with oxygen breathing (a 400% increase in oxygen tension) and a greater fall in saturation with exercise. Recognizing that normally the right lung contributes some 55% of the ventilatory function, the bronchospirometry clearly indicated that the mechanical problem was essentially in the right lung. Nevertheless, the right lung was taking up oxygen well in proportion to its ventilation, suggesting that its problem was chiefly mechanical.

The esophageal pressure studies provided the direct confirmation of the patient's functional problem. In the expiratory position, there was a pressure of 4 cm H_2O across the lung. With the addition of 500 ml of air and the relaxation of the respiratory muscles, there was a pressure of $9 - (-7)$ or 16 cm H_2O across the lung. The volume increase of 500 ml was, therefore, associated with an increased pressure across the lung of $16 - 4 = 12$ cm H_2O. This yields a pulmonary compliance of 0.04 liters/cm H_2O, compared with the normal value of 0.20 liters/cm H_2O. It is interesting that to adapt to this low pulmonary compliance and the large amount of work necessary to stretch this restricted lung, the patient had adjusted to relatively rapid but shallow tidal respiration (compare with Case 2).

The laboratory data encouraged the surgeon to remove the fibrous scar encasing the right lung, and the patient's respiratory problems were greatly alleviated.

4

A 28-year-old man was referred to a pulmonary function laboratory because of extreme dyspnea that required constant oxygen therapy. At the age of 23 he had gone to work in a chemical plant; he was employed there for three years until he quit his job because he heard rumors that there was dangerous contamination of the air in the plant. After about six months, during which he found temporary employment at odd jobs, he began to note shortness of breath with exertion. He was examined by his physician, who found no evidence from physical examination or x-ray to explain his respiratory problem. During one of his extreme bouts of dyspnea, he was admitted to the emergency room of his local hospital, where he was given oxygen therapy. His dyspnea promptly subsided. His condition had progressed to such a point that he had to be maintained with almost constant oxygen therapy to avoid debilitating dyspnea. The patient also reported that he had experienced several attacks of convulsive seizures.

It was necessary to carry out the ventilatory tests with oxygen. Vital capacity was 3.66 liters, 80% of expected normal. Maximum ventilatory volume was 99.4 liters/min, 68% of expected normal. His forced expiratory volume in 1 second was 86% of his vital capacity. Arterial oxygen saturation was estimated from oximetry to be 98.5% while breathing room air. The patient was then connected to an air-filled closed-circuit spirometer with CO_2 absorption, yielding a respiratory tracing with an upward drift of the base line as a function of oxygen consumption. The spirometer was then flushed thoroughly from an oxygen tank and another recording was obtained. The spirograms are shown in Figure 4-1. The spirogram reads from right to left, with time on the abscissa and tidal volume on the ordinate, the rising excursions representing inspiration. Because of the difficulties of working with this patient, no further tests were done. The patient was referred back to his physician with an evaluation of his problem.

Question
What could account for the respiratory pattern shown in the spirogram?

Analysis of Case 4
The conspicuous feature in the history of this case is that there had never been any report of cyanosis in spite of this patient's dependence on oxygen. Yet his physician, with the techniques he had available,

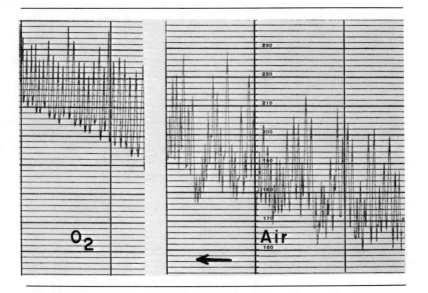

Figure 4-1. Respiratory tracing of the patient breathing room air and breathing from an "oxygen" tank. The tracing reads from right to left with inspiration up.

could not completely discount the alleged history of exposure to a toxic atmosphere.

In view of the difficulties of carrying out ventilatory studies in a patient so severely debilitated, the minor deficiencies noted were disregarded, and his pulmonary mechanics were assessed as excellent. The fact that his arterial saturation was so high while breathing room air argued against a significant diffusion defect and suggested that the patient was hyperventilating. Bouts of respiratory alkalosis associated with periods of hyperventilation can lead to tetany and could have accounted for the report of "convulsive seizures."

The nature of this man's problem is demonstrated in the spirogram. The bizarre respiratory pattern observed on air breathing, with its great irregularity in both rate and depth, does not accord with the functional properties of the medullary respiratory center, which normally maintains the smooth respiratory pattern of eupnea. Particularly noteworthy are the highly irregular expiratory positions at the bottom of the trace. Normal expiration is an essentially passive process, during which elastic forces in the chest wall and lung come into an equilibrium position that shows only minor variations from one breath to another. The disappearance of this chaotic respiratory pattern on the administration of oxygen also does not accord with the known physiological responses of the respiratory system. Actually, the description of this test was intentionally deceptive in stating that the spirometer was flushed from an oxygen tank before recording the O_2 segment of the spirogram. The tank employed was conspicuously labeled "oxygen," but its oxygen contents had long since been exhausted and it had been recharged from the compressed air line; it was kept in a corner of the laboratory for just such purposes.

This patient was, therefore, evaluated as suffering from a respiratory neurosis. Higher brain centers were driving the respiratory motor neurons, presumably by way of the corticospinal system, and bypassing the medullary centers normally responsible for respiratory control. Readers unfamiliar with this type of case should be cautioned against dismissing this problem as trivial. The patient was just as severely crippled by this functional disturbance as he would have been by severe organic lung disease. The value of the laboratory evaluation was to provide a secure basis for assuring the patient that he should seek psychotherapy to resolve his respiratory difficulties.

5

A young mother was admitted to the hospital in coma. Her husband reported that prior to his discovering her in an unconscious state, she had been in good health except for complaints of nervousness and excessive fatigue. Examination revealed a pulse of 122 beats/min, blood pressure of 100/50, shallow respiration of 9 per minute, and a warm but pale skin. The rest of her physical examination revealed nothing of significance. An analysis of her arterial blood demonstrated 9 gm of hemoglobin with an oxygen content of 10 vol% and HCO_3 of 27 mM with a plasma H^+ of 48 nM. Her RBC count was 4.1 million/mm^3

By the time the blood analysis had been obtained, her respiration had changed to periods of apnea, alternating with periods of dyspnea. She was placed on oxygen and an additional chemical test was run on her blood. A special procedure was then instituted.

Questions

On the basis of her initial examination, without regard to her blood analysis, what types of physiological disturbances can you identify? Can you identify any common physiological factor that could explain the whole picture?

What would you estimate her arterial PO_2 to be? What would be the most likely cause of her hemoglobin abnormality? How do these factors relate to her condition?

Calculate her PCO_2 and explain the acid-base disturbance that she represents.

Explain the physiology of the disturbed respiration that she exhibited before oxygen therapy was instituted. Should oxygen have been given to this patient?

This problem is not uncommon in modern clinical practice and is often observed in animal laboratory experiments. What should have been looked for in her blood? If, as in this case, its concentration was life-threatening, how could it be removed from her blood?

Analysis of Case 5

This woman displayed the signs of severe depression of central nervous system function, involving not only the sensorium but brain stem function as well. The wide pulse pressure, with a low diastolic pressure in spite of the tachycardia, together with the warm skin, indicate a generalized vasodilation that suggests depression of sympathetic tone. In this situation, the tachycardia would indicate depressed vagal tone. The breathing observed on admission was clearly that of respiratory-center depression. Depth and rate of respiration are so highly variable that neither factor in itself is of great significance until very abnormal values are observed, but the combination of a slow rate of respiration with a shallow depth is ominous.

A pale skin signifies either anemia or a state of vasoconstriction; the fact that this patient's skin was warm led to the prompt recognition of her anemia. The fact that the hemoglobin was depressed significantly more than the red blood cell count identifies the anemia as the hypochromic type, presumably related to iron deficiency, which is a common occurrence following pregnancy. This anemia could have contributed to her cardiovascular status of a high-output vasodilated state, but it is not nearly of the severity that would be required to account for her depressed neural function. An arterial oxygen content of 10 vol% with 9 gm of hemoglobin represents a saturation of 83%, which should correspond with a PO_2 of about 50 mm Hg with the acidosis that was present. Since the cerebral vasculature is well autoregulated to protect against hypoxia, especially in the presence of elevated PCO_2, the patient was comfortably above the level where a significant impairment in oxygen delivery to the brain would be encountered (cf. Case 9). Her PCO_2 is calculated as 54 mm Hg, confirming the presence of a respiratory acidosis. Although an extreme respiratory acidosis can produce coma, the PCO_2 in this instance is far below the level of CO_2 narcosis.

The appearance of Cheyne-Stokes respiration was further evidence of severe depression of brain stem function and could also denote further deterioration of the circulation. With a prolonged circulation time and depressed responsiveness of the respiratory centers, smooth regulatory control of the respiratory system breaks down into pendular swings of the feedback control loops that oscillate the system between dyspnea and apnea. Oxygen therapy was of value to minimize the superposition of hypoxic depression on the preexisting depression of her respiratory centers during the apneic phases of the Cheyne-Stokes breathing.

Barbiturates are widely used drugs that can produce this pattern of central nervous system depression. A blood test for barbiturates in this patient revealed a very high blood level. The patient was judged to be so critical that it was not felt prudent to wait for the liver to metabolize the drug, and, therefore, the patient was given an emergency hemodialysis to lower the blood level of the drug. On recovery, the patient reported that she had become so desperate to obtain some rest from her domestic responsibilities that she had ingested an indiscriminate number of sleeping pills with the hope of catching a good nap before the baby awoke. Her new chores of motherhood, coupled with her high susceptibility to fatigue, because of her anemia, had led her to this nearly fatal indiscretion.

A 59-year-old carpenter entered the emergency room complaining of shortness of breath. His pulse rate was 112 beats/min with a blood pressure of 138/88. His respirations were 35 per minute and extremely labored. Loud breath sounds with some crackling rales were heard over all lung fields. The patient had a rather ashen complexion and his nail beds gave clear evidence of cyanosis. An arterial blood sample was obtained, a chest x-ray was ordered, and the patient was placed in an oxygen tent.

One-half hour later the physician was called to the bedside by the nurse, who found the patient unresponsive. The patient's complexion had changed to a flushed pink with no trace of cyanosis. His respirations were quiet at a rate of 10 per minute. His heart rate was 140 beats/min with a blood pressure of 85/50. The patient was in deep coma.

A call to the laboratory to check the blood tests revealed a hemoglobin of 16 gm with a hematocrit of 52%, an arterial PO_2 of 48, a PCO_2 of 90, an H^+ of 45 nM, and an HCO_3 of 48 mM.

Questions
Explain the transformation of the patient's status after his admission to the hospital.

What would you predict his x-ray would show?

What type of therapy should such a patient be given?

Analysis of Case 6
This is a case of CO_2 narcosis produced by inappropriate oxygen therapy. The blood chemistry demonstrated a very severe respiratory acidosis that was reasonably well compensated; his actual acidemia (H^+ increase) was slight because of the large amount of bicarbonate retention by the kidneys. This high bicarbonate, together with the evidence of a compensatory polycythemia, identifies chronic respiratory insufficiency of a severe degree.

The CO_2 retention associated with chronic respiratory insufficiency tends to make the fluids in the brain acidotic, as evident from the relationship:

$$H^+ = 24 \ PCO_2/HCO_3^-.$$

To protect the brain from acidosis, there are metabolic "pumps" capable of translocating bicarbonate across the blood-brain barrier into the cerebrospinal fluid. It must be appreciated, however, that the central chemoreceptors in the brain stem that adjust the respiratory activity to control the PCO_2 are actually stimulated by the concentration of H^+ in the receptor. Referring back to the Henderson equation given above, note that as the HCO_3^- in the cerebrospinal fluid increases, it will require progressively greater increases in the PCO_2 to produce a given increase in H^+. This relationship is of no consequence in the normal moment-to-moment regulation of the PCO_2 because it requires several hours for the pumps to achieve any significant change in bicarbonate concentration. In a chronic disturbance, however, this mechanism for protecting the brain against acidosis has the unfortunate consequence of lowering the sensitivity of the respiratory system to CO_2.

Fortunately nature has evolved an alternate mechanism for maintaining the respiratory drive. As the degree of chronic respiratory insufficiency becomes greater and the responsiveness of the central chemoreceptors to CO_2 consequently reduced, arterial oxygen tensions fall into the range where the peripheral chemoreceptors in the carotid and aortic bodies are strongly stimulated. This situation was clearly present in this patient, with the arterial PO_2 close to the maximal level for hypoxic stimulation of the peripheral chemoreceptors, while the PCO_2 had risen close to the range where CO_2 becomes a depressant rather than a stimulant to respiration.

Placing such a patient in an atmosphere that would run in excess of 50% oxygen, or an inspired PO_2 on the order of 400 mm Hg, would raise his arterial PO_2 to well over 100 mm Hg in spite of his pulmonary insufficiency. This would have shut off his hypoxic respiratory drive. The consequent fall in pulmonary ventilation would produce still further elevation of the PCO_2 into the range where CO_2 becomes a narcotic. These very high CO_2 levels also act on the peripheral tissues to produce a generalized vasodilation and compromise circulatory function. Since renal compensation can not function rapidly to combat an acute acidosis, the patient's life was imperiled.

It is important to identify patients with this severe a degree of respiratory insufficiency and be very conservative in the administration of oxygen. A modest enrichment of their inspired oxygen tension via a nasal catheter will protect the patient from severe hypoxia while the patient is given a careful work-up to determine whether any other steps can be taken to improve his respiratory status.

Patients suffering from pulmonary insufficiency of this severity usually have very advanced emphysema, characterized on x-ray by very radiolucent lung fields with scant evidence of the normal fine markings of lung tissue. Evidence of bronchitis would be important, since it would represent a treatable complication that might have been responsible for precipitating this acute exacerbation of this patient's problem.

Blood

11

A 3-year-old Puerto Rican boy was brought to the emergency room by his mother. It was difficult to obtain the history because there was no one immediately available who spoke Spanish, but it appeared that the immediate cause for bringing the child to the hospital had been a couple of fainting spells. The child gestured, indicating that his head ached. His blood pressure was 62/48 with a pulse rate of 132 beats/min. His respiratory rate was 28 per minute and somewhat labored. The chest examination was essentially negative; abdominal examination revealed diffuse tenderness and very active bowel sounds.

It became apparent to the examiner that the patient's skin color did not represent normal ethnic pigmentation but, rather, that the child was deeply cyanotic. An arterial blood sample was obtained for laboratory analysis and the child was given oxygen therapy. With oxygen therapy his respiration appeared to quiet somewhat, but his deep cyanosis persisted. The electrocardiogram was interpreted as showing a sinus tachycardia with normal complexes.

Blood values were reported back as a PO_2 of 116 and an arterial oxygen content of 9 vol%, with a hemoglobin of 16.4 gm. At about this point the child passed a loose stool that was noted to be black in color. A further chemical test on the blood was carried out and specific therapy was instituted. Within 4 hr the cyanosis had disappeared and the child's vital signs returned to within normal limits. The mother was instructed on the appropriate care of the child and the patient was discharged.

Questions
What are possible causes of cyanosis? What can be concluded from the failure of oxygen therapy to relieve the cyanosis?

What is the cardiovascular status of this patient? Is a shocklike state to be anticipated in a case of congenital cyanotic heart disease?

Could hemorrhage into the gastrointestinal tract explain this picture?

Analysis of Case 7
Acute respiratory problems severe enough to cause cyanosis would scarcely be compatible with a negative chest examination, and most pulmonary causes of inadequate oxygenation of the arterial blood would show a relief of the cyanosis by the very large increase in oxygen tension resulting from oxygen therapy.

The shocklike picture of the patient is not characteristic of cyanotic heart disease in that the latter would be expected to demonstrate a compensatory increase in cardiac output and a wide pulse pressure. Beyond that, this child did not present a picture of shock. A patient in shock will show intense cutaneous vasoconstriction and, hence, not demonstrate deep cyanosis. While the black stool could be suggestive of GI bleeding, this should be associated with compensatory vasoconstriction and mobilization of tissue fluid into the plasma compartment, with a consequent dilution of the hemoglobin.

The blood data clearly identify an abnormality of the hemoglobin in that there was a significant decrease in the arterial oxygen content in spite of a high PO_2 (attributable to the labored respiration) and the high hemoglobin (suggestive of dehydration).

Laboratory analysis confirmed that some 55% of this child's hemoglobin was in the form of methemoglobin. Arrival of a Spanish-speaking member of the hospital staff clarified that the child had been suffering from diarrhea for the previous three days, which his mother had been treating with bismuth subnitrate, which accounted for the black coloration of the stool. Apparently the intestinal flora in this child had been rapidly converting the nitrate to nitrite, which was accounting for the methemoglobin formation. Nitrite is also a potent dilator of the vascular bed, accounting for the low blood pressure, the fainting spells, and the headache. Treatment of this case employed oral methylene blue, which served as an oxidizing agent to convert the methemoglobin back into the oxidized form and, obviously, the substitution of other therapy for alleviation of the child's diarrhea.

A secretary who was working for a firm in Brooklýn decided to spend her vacation in Mexico City. A few weeks after her return she visited her doctor complaining of nocturia that was disturbing her sleep. She had no other complaints, although admitted that perhaps she had been drinking more water since her return from her vacation.

Examination of the urine revealed a dilute urine that was free of glucose and protein. Physical examination revealed nothing else of significance. Her physician made another appointment for her and requested that she drink as little fluid as possible during the 18 hours preceding that appointment. On the second visit the patient complained bitterly of extreme thirst. Urinary examination revealed a dilute urine with no other significant findings.

The patient had an older brother who had received extensive treatment for complications of sickle cell anemia.

Questions

What is the immediate difficulty responsible for the nocturia?

What does the family history identify as the underlying cause?

Why did the vacation in Mexico City precipitate this problem? (Mexico City is at an altitude of 7350 ft).

Analysis of Case 8

Nocturia is most commonly the consequence of an increased urine volume. This could be directly caused by a primary increase in water intake and, in all cases, will be associated with a secondary increase in water intake to preserve body water balance. Other causes of increased urine flow would be an increase in the osmotically active material being excreted by the kidney or a failure of renal mechanisms to concentrate the urine. In this patient the absence of glucose in the urine excluded the most common cause of osmotic diuresis, and the persistence of a dilute urine in the face of water deprivation established a failure of the concentrating mechanism of the kidney as the immediate cause of this young woman's problem.

The family history identifies this patient as black and suffering from a disease that tragically afflicts many of her race. The abnormal hemoglobin in sickle cell disease has altered solubility properties that result in the hemoglobin precipitating out of solution in the presence of (1) reduced oxygen tension, and (2) increased osmolarity. These rigid precipitates distort the red blood cells into the sickle cell configuration, and the distorted rigid cells tend to stick and clump together and block the microcirculation. In the homozygous form of the disease, which presumably was represented in the brother of this patient, the tendency to sickling is high, and patients encounter multiple problems that result from blockage of the microcirculation at sites throughout the body. The patient with the heterozygous trait and, consequently, a lesser amount of the abnormal hemoglobin has cells that are much less likely to sickle and, therefore, escapes the ravages of the disease.

There is one site in the body, however, that is uniquely susceptible to sickle cell damage. The loop of Henle in the renal medulla forms a countercurrent multiplier that generates a reservoir of high osmotic concentration, which, in the presence of antidiuretic hormone, provides a steep osmotic gradient for extracting water from the renal collecting ducts to concentrate the urine. The medullary pyramids that contain this osmotic reservoir are supplied with blood by vasa recta, which have a relatively sluggish flow of blood to avoid washing out the osmotic gradient. Red cells traversing the vasa recta are, therefore, exposed to an environment with a high osmotic pressure and a rather low oxygen tension. The visit to Mexico City would have accentuated both these factors. Compared with the inspired oxygen tension of approximately 159 mm Hg that this secretary encountered in Brooklyn, at the altitude of Mexico City she was exposed to an inspired oxygen tension of 122 mm Hg. This would lower her normal arterial PO_2 of 100 mm Hg down to 72 mm Hg. At the same time, American tourists are strongly cautioned about problems with unsafe drinking water in Mexico, and their relative abstinence from fluid intake can make them somewhat dehydrated. These factors apparently summated in this patient to produce enough sickling to block the medullary vasa recta and destroy the renal concentrating mechanism in the medullary pyramids of her kidneys. Without the high osmotic gradient created by the ionic pumps in the thick segment of the ascending limbs of Henle's loops, there is no osmotic reservoir to extract water from the collecting ducts when the water permeability of the ducts is increased by the action of the antidiuretic hormone.

It should be recognized that this patient will compensate for this deficiency by relying on her thirst mechanism to maintain fluid balance. Learning to live with some chronic polyuria is a far less drastic consequence of this disease than occurs with the diffuse obstructions to other areas of the vascular bed observed in patients with the homozygous disease.

A 44-year-old textile worker was in excellent health until the day before Thanksgiving, when he noted a runny nose and a headache and assumed he was catching a cold. He dosed himself with aspirin and nose drops. On Thanksgiving he joined in family festivities and ate a good dinner but felt very fatigued. That evening his wife noted a yellow coloration of his skin. The next morning he had breakfast and started out for work, but he returned home and went to bed when he became overwhelmed with fatigue. By evening he complained of feeling light-headed and exhibited emotional instability and irrational behavior. He was admitted to a local hospital for observation. Their findings confirmed severe jaundice, anemia, and hemoglobinuria. The next morning he was transferred to a medical center.

On admission to the medical center he was very jaundiced and in a confused and mentally disoriented state. His blood pressure was 115/60, his pulse was 120 beats/min, and his respiration was 14 per minute. When a hematocrit of 11% was found, blood was sent to the laboratory for further analysis, including a Coombs' test, and he was given 1 unit of packed red cells. When the Coombs' test provided positive evidence that he was hemolyzing his red cells, it was elected to withhold further transfusions, and he was treated with a broad-spectrum antibiotic for a possible infectious agent.

By the middle of that afternoon the patient had lapsed into coma. Neurological examination revealed a Babinski reflex, widely dilated pupils that were unresponsive to light, and outward rotation of both eyeballs. An EEG showed minimal activity. Packed red cells were administered intra-arterially under pressure while the EEG was monitored. EEG activity was restored almost immediately with the administration of cells, and within a couple of minutes the patient had regained consciousness with a clear and alert sensorium. During the following 48 hr, several additional units of cells were administered with the aim of maintaining the hemoglobin level above 5 gm/100 ml. The patient progressed well, evidence of hemolysis progressively disappeared, and his urine cleared. On his third hospital day, a creatinine blood level of 0.8 mg/100 ml and a urine osmolality of 585 mOsm were determined. Laboratory data on his first day in the hospital are shown in the table.

Questions
What is the explanation of this patient's jaundice?

Blood chemistry	Admission	With coma
H^+ (nM)	36	36
HCO_3^- (mM)	17	9
PCO_2 (mm Hg)	25	20
Hemoglobin (gm/100 ml)	3.4	3.2
Hematocrit (%)	11	10
Total bilirubin (mg/100 ml)	11.7	
Conjugated, direct (mg/100 ml)	3.3	
Unconjugated, indirect (mg/100 ml)	8.4	

Why was the patient's bicarbonate depressed?

How do you reconcile this depressed bicarbonate with a slightly low hydrogen-ion concentration?

What accounted for the neurological findings?

How could a small increase in this patient's red cells produce such a dramatic restoration of cerebral function?

Of what significance were the urinary findings on his third hospital day?

Analysis of Case 9
This patient exhibited a dramatic hemolytic crisis as evidenced by the very low hemoglobin that continued to drop in spite of the administered cells and by the very high bilirubin levels causing the jaundice. Degradation of the liberated hemoglobin flooded the system with bilirubin, which has a low aqueous solubility and is carried in the plasma bound to protein. Because of this protein binding, it can only be measured indirectly in the laboratory by treating the sample with alcohol. To prepare the bilirubin for excretion, the liver solubilizes the bilirubin by conjugating it, chiefly with glucuronic acid, and this material reacts directly in the laboratory test. The data for this patient show a conjugated bilirubin level that is far above normal, suggesting good liver function, even though this liver function was being overwhelmed by an even greater supply of the unconjugated protein-bound bilirubin being produced by hemoglobin degradation. Further studies of fecal and urinary products of bile pigment metabolism could have confirmed that impaired excretion of bile was not contributing to the jaundice, but additional studies to explain the jaundice are scarcely warranted in a patient with such unequivocal evidence of massive hemolysis.

The impaired oxygen delivery to the tissues that was associated with the anemia forced cellular metabolism to resort to anaerobic pathways to obtain energy, with a consequent formation of increased lactic acid. The increase in lactic acid accounts for the extreme fatigue experienced by this patient, and, as the deficiency became more severe, lactic acid production led to depletion of the bicarbonate buffer and a metabolic acidosis.

In spite of this metabolic acidosis, hydrogen ion was a bit low and the CO_2 tension was quite low in this patient, indicating a superimposed respiratory alkalosis. This was the result of chemoreceptor activation. The arterial chemoreceptors contain a special cytochrome that has a much lower affinity for oxygen than the typical mitochondrial cytochrome, which makes the chemoreceptors sensitive to small reductions in their local tissue oxygen tension. Chemoreceptor tissue has a very high rate of blood flow that quickly replenishes oxygen consumed by the receptor cells themselves. Therefore in most situations the oxygen tension in these receptor cells is almost exclusively determined by the oxygen tension in the arterial blood. This arterial oxygen tension is the result of equilibration of gas tensions in the pulmonary capillaries, a physical process that is independent of the amount of hemoglobin in the blood. With mild to moderate degrees of anemia, therefore, there is no hypoxic signal to stimulate the chemoreceptors, and shortness of breath is not a characteristic complaint of the anemic patient. With severe states of anemia, the reservoir of oxygen within the red cells becomes so limited that the metabolism of the chemoreceptor tissue does tend to deplete the supply and lower the local tissue tension. This action stimulates the receptors and increases the respiratory drive. The increased respiration will raise arterial oxygen tension, but this can do little to increase the oxygen supply being delivered by the limited number of red cells available, since they are already nearly saturated.

The Coombs' test identified this hemolytic crisis as an autoimmune disease in which the patient was destroying his red blood cells at a rapid rate. This process released hemoglobin into the blood at a much faster rate than it could be metabolized. Since free hemoglobin in the plasma is filtered by the renal glomeruli and becomes concentrated in the renal tubules, this hemoglobin can cause serious damage to the kidneys (cf. Case 27). The desirability of restoring the red blood cell concentration, therefore, had to be weighed against the possibility of producing renal damage by the administration of more red cells.

The neurological findings that appeared during the afternoon indicated grave depression of brain function. Comas are the result of depression of the cortical circuits that are responsible for consciousness but do not usually involve the subcortical circuits that are responsible for the basic rhythms observed in the electroencephalogram. A "flat" EEG is, therefore, a very ominous sign of total cerebral depression. This evidence was correlated in this case with the appearance of a positive Babinski sign, signifying loss of pyramidal-tract control of spinal motor systems, and by the eye signs that resulted from depression of mid-brain circuits. This depression of brain function was due to a critical lack of oxygen.

The brain normally removes about 50% of the oxygen delivered to it in its arterial blood supply. When this blood lacks an adequate oxygen content, autoregulatory compensation can dilate the cerebral vasculature so as to approximately double its blood flow. This represents a fourfold factor of safety, whereby brain function can be maintained when the arterial oxygen content is reduced to as little as one-quarter of its normal value of 20 vol%, or 5 vol%. With a hemoglobin of only 3.2 gm/100 ml, this patient's arterial oxygen content would have dropped to $3.2 \times 1.34 = 4.3$ vol%, below the minimal level for compensation. This level would jeopardize the maintenance of brain function and also cardiac function (cf. Case 10). In this patient, the functional failure was observed most conspicuously in the brain, probably because his acutely lowered CO_2 tension would inhibit the maximal dilator response of the cerebral vasculature. Therefore inadequate oxygen was being supplied to maintain cerebral function. Yet a relatively small increase in the oxygen-carrying capacity of the blood would have elevated the oxygen supply above this minimal threshold to restore cerebral activity, as was demonstrated by the dramatic return of electro-encephalographic activity and then consciousness as this patient was rapidly transfused with cells. The goal of repeating these transfusions as necessary to maintain a hemoglobin level of over 5 gm/100 ml was equivalent to maintaining an arterial oxygen content of over 6.7 vol%, which should have provided the necessary oxygen supply for brain function with a small margin of safety.

As stated above, a more aggressive treatment of the anemia would have run the risk of producing renal damage. The urinary findings on the third hospital day indicated that the patient had the good fortune to escape this hazard of the disease. Not only was there an absence of creatinine retention due to obstructed glomerular filtration, but the ability of the kidney to concentrate the urine — one of the most vulnerable of renal functions to damage from hemoglobinuria — had been preserved.

The causes of these transient episodes of autohemolysis are

poorly understood. They are inappropriate responses of the immune system, presumably triggered in susceptible patients by the invasion of the body by certain infectious agents. It will be recalled that this patient had minimal signs of an upper respiratory infection preceding his hemolytic disease; whether this infection was the agent that evoked this reaction is a matter for speculation.

A 72-year-old widow was brought to the hospital after a neighbor found her collapsed on the floor of her apartment. She was very pale, extremely weak, and complained of fatigue. She was very emotionally depressed. She responded rather sluggishly but reasonably coherently to questioning. She explained a stain down the front of her dress as the result of spilling a cup of beef broth that she had been too weak to hold.

Her blood pressure was 88/62, her pulse rate was 56 beats/min, and her body temperature was 30.0°C. Her respiratory rate was 22 per minute and labored. Auscultation of the chest revealed congestion at both lung bases and an enlarged heart. She was admitted to the intensive care unit, and a venous catheter was passed and wedged into her pulmonary vasculature to provide an estimate of left atrial pressure, which was found to be 16 mm Hg. Admission laboratory data were as follows:

Na	149 mM	Cl	84 mM
K	4.9 mM	HCO_3	3.6 mM
H^+	66 nM	Lactate	64 mM
PO_2	126 mm Hg	Hemoglobin	2.7 gm/100 ml
PCO_2	10.1 mm Hg	Hematocrit	6.8 %

Over the next 12 hours the patient was transfused with packed red cells, which raised her hemoglobin of 6.1 gm/100 ml and her hematocrit to 18%. She responded to this treatment with a rise in blood pressure to 130/55 and a pulse rate of 74 beats/min, and her body temperature rose to 36.1°C. At this time her arterial PO_2 was 89, PCO_2 was 35.2, H^+ was 44 nM, and bicarbonate was 19.3 mM; wedge pressure was 8 mm Hg.

Though still emotionally depressed, the patient reported that she felt much better. Further questioning brought out that economic inflation had made it difficult for her to survive on her meager income and her pride would not permit her to seek welfare assistance; consequently, for several months her diet had consisted almost exclusively of beef broth and soda-pop.

Questions

To understand the acute problem in this patient, estimate her arterial oxygen content at the time of admission, and relate this to the clinical findings.

How do you account for the dramatic response to the transfusion of red blood cells?

Why was the patient given packed cells rather than whole blood?

What is the most obvious explanation for this type of anemia?

Analysis of Case 10

This extraordinary degree of anemia of a microcytic type obviously related to this patient's severe malnutrition and iron deficiency. With a hemoglobin of 2.7 gm/100 ml, her arterial oxygen content would have been only $2.7 \times 1.34 = 3.62$ vol%, which is less than half the volume of oxygen usually extracted from the coronary blood. Even with autoregulatory dilation of the coronary bed, this volume would not provide enough oxygen to sustain cardiac action, especially with the high cardiac output required to support body metabolism in the presence of anemia. On admission to the hospital she was, therefore, in severe myocardial failure due to severe anemic hypoxia, as indicated by the narrow pulse pressure in spite of the high left atrial pressure and the prolonged filling time of the ventricle associated with the bradycardia. The lactic acid acidosis was due to this anemic hypoxia, which was augmented by a stagnant hypoxia that resulted from her myocardial failure.

With an increase in her red cell mass, arterial oxygen content would have risen to $6.1 \times 1.34 = 8.17$, restoring an adequate supply of oxygen to the myocardium. Thus there was a remarkable improvement in myocardial function and cardiac output, as indicated by the wide pulse pressure and the fall in wedge pressure. The low diastolic pressure after treatment reflects the large blood flow into the peripheral bed dilated by its metabolic needs. With this restoration of oxygen supply and blood flow, the lactic acid that had accumulated was substantially metabolized to correct the acidosis.

Transfusion of a patient in heart failure always represents a dilemma to the physician, since the decompensated left ventricle is failing to increase cardiac output to keep pace with the venous return to the heart. Adding further fluid volume to the venous return may merely aggravate the ventricular failure and increase pulmonary congestion, causing pulmonary edema, which is the immediate life-threatening complication of left ventricular failure. Packed red cells provide the patient with the urgently needed oxygen-carrying capacity without the addition of any further volume load on the failing heart.

An important feature of this case history was the slow chronic course of many months, during which the anemia developed. (Contrast this case with Case 9.) As a general principle, it is known that the enzymatic machinery of the body can adapt to chronic metabolic stresses and, specifically, that a stabilized condition of hypometabolism can develop with chronic undernutrition. This woman encouraged that adaptation on three counts: deficient calorie intake, deficient intake of vitamins and other cofactors to utilize calories, and a progressive anemic hypoxia that interfered with the metabolism of those calories. The hypometabolic state that this patient had undoubtedly developed would result in less heat production. Associated with this, she developed hypothermia. This lowering of body temperature would decrease metabolic activity and lead to further hypometabolism. By this type of feedback loop, this patient was able to survive and even maintain consciousness, in spite of the extreme anemia that she developed.

Cardiovascular System

III

A 27-year-old unemployed male appeared at the clinic complaining of throbbing headaches. His blood pressure was 184/52, his pulse rate was 70 beats/min, and chest examination demonstrated an enlarged heart. On auscultation, a high-pitched systolic murmur and an accentuated second heart sound were noted. In completing the physical examination, a scar was noted on the inner aspect of his thigh. Palpation revealed a strong thrill in the area of this scar. The patient initially dismissed this scar as an "accident." On closer questioning, he finally admitted that it was the result of a gun-shot wound. The bullet had not lodged in the tissue, and he had been able to treat the wound with first-aid measures. He had not sought medical assistance because he did not wish to reveal the circumstances under which he received the wound.

The patient was admitted to the surgical service.

Questions
What is the nature of the patient's problem?

Why would a lesion of this nature manifest itself to the patient in the form of a headache?

Analysis of Case 11
The very wide pulse pressure with a low diastolic value indicates that this patient's heart was pumping a very large stroke volume associated with a low resistance pathway for diastolic runoff. The thrill over the scar on the thigh suggests that this low resistance pathway is a large fistula, connecting between the femoral artery and the femoral vein. A large flow through this A-V shunt returns a significant increment of blood volume to the heart to augment the normal venous return.

The fact that the original wound could be effectively treated by first-aid measures indicates that there was no massive hemorrhage. Nevertheless, there must have been severe trauma to the blood vessels adjacent to the path of the bullet, with consequent necrosis and regeneration of the elements of the vascular wall. It is an interesting morphological phenomenon that vascular cells recognize other vascular cells in the process of growth and regeneration and are not always selective as to whether they are arterial or venous cells. In the architectural reorganization of the necrotic tissue that resulted from the gun-shot wound, arterial elements had become associated with venous elements, reconstructing a structurally sound but anatomically incorrect continuity of the vascular wall between a large artery and a large vein.

The cardiac systolic murmur noted was undoubtedly in the category of a "functional" murmur. In a normal heart, the ejection velocity of blood into the ascending aorta is not too far below the critical Reynold's number, above which it becomes impossible to sustain uniform laminar flow of the blood stream. With the greatly increased venous return and diastolic distention of the left ventricle, a vigorous Frank-Starling response of the myocardium creates a high ejection velocity to significantly exceed the Reynold's number. This creates turbulent eddies that generate local vibrations of the vascular wall, which are responsible for the murmur. The same basic phenomenon was also responsible for the thrill vibrations detected over the arteriovenous fistula. the aortic valve as an alternative diagnosis, the fact that the cardiac murmur occurred during systole and that the second heart sound was accentuated would argue against such a diagnosis.

Most visceral tissue appears to lack any specific receptor system that is capable of eliciting a pain sensation. A notable exception to this generalization is blood vessels, which evoke strong pain sensations when distended excessively. This mechanism is especially well developed in the extracranial vessels. The very wide pulse pressure was producing excessive distention of these blood vessels with each pulse and thereby serving as the source for the throbbing headache.

12

A 22-year-old salesman was discovered to have high blood pressure on a routine physical examination and was referred for a complete workup. He reported that he considered himself to be in excellent health. On closer questioning, he admitted to occasional throbbing headaches and was annoyed by cold feet in the winter months. His resting pulse rate was 70 beats/min. Auscultation of the chest indicated a somewhat enlarged heart. Measurement of his blood pressure in the usual fashion demonstrated a pressure in the brachial artery of 168/120. When attempts to find pulsations in the dorsalis pedis artery were unsuccessful, blood pressure was measured in the femoral artery and found to be 70/50. Special x-ray examinations were then carried out to confirm the suspected diagnosis, and the patient was admitted for corrective surgery. One week postoperatively, his pressures were 115/76 in the brachial artery and 120/74 in the femoral artery. On a follow-up examination three months later, the patient appeared to be in good health. His brachial pressure at that time was 145/98 and his femoral pressure was 152/96.

The patient was next seen seven years later, when he was admitted to the hospital with severe dyspnea. Chest examination revealed a greatly enlarged heart and pulmonary congestion. His brachial blood pressure was 175/128 and his femoral blood pressure was 186/125. His liver was significantly enlarged, and there was considerable ankle edema.

He was treated with oxygen, a cardiac drug, and an antimetabolite of aldosterone. For three days he showed considerable improvement but on the fourth day he complained of diarrhea. On the afternoon of the fourth day the nurse found him collapsed on the floor of the bathroom complaining that he was too weak to get back to his bed. She summoned an orderly and they returned the patient to his bed; he exhibited such profound muscular weakness that he could offer little assistance in the maneuver. His blood pressure at this time was 114/78. His hematocrit was 46%. Neurological examination was negative except for generalized muscular weakness with minimal facial expression and slow eye movements. The patient's blood chemistries at this time were as follows:

Na	132 mM	Cl	106 mM
K	6.8 mM	HCO_3	17 mM
Glucose	85 mg/100 ml	Creatinine	3.2 mg/100 ml
BUN	41 mg/100 ml		

Questions

What was the nature of the original disease in this patient?

What was the nature of his problem on his second hospital admission, and how might it have related to his original difficulty?

What was responsible for the complication that developed in this patient on his fourth day in the hospital?

Analysis of Case 12

The original picture presented by this young man was one of diastolic hypertension, indicating an abnormally high resistance to aortic blood flow. The hypotension in the lower extremities indicated that this resistance was localized in the central aortic conduit, pointing to the diagnosis of coarctation of the aorta, due to improper fusion of the embryonic arches. It should be especially noted that, except for the rather trivial complaint of cold feet, there was no evidence of any significant deficiency in blood flow to regions of the body below the coarctation. Thus his initial femoral pressure of 70/50 was still within the range where autoregulatory mechanisms at the microcirculatory level of the peripheral tissues should be able to compensate for the deficiency in the arterial pressure and maintain adequate blood flow. Appropriate surgery removed the coarctation and eliminated the necessity for this compensatory mechanism to maintain blood flow to the lower extremities.

Seven years later the patient returned to the hospital in congestive heart failure that was secondary to a return of his hypertension. The hypertension is now generalized and would fall in the category of *essential* hypertension. Although explanations of the etiology of almost any form of essential hypertension are highly speculative, this particular form lends itself to a plausible hypothesis that can be reproduced in animal models. For 22 years, the coarctation had caused a chronically low renal arterial blood pressure. This lowered renal perfusion pressure would have been expected to activate the renin-angiotensin-aldosterone system to increase vascular tone and foster fluid retention to maintain an augmented blood volume. These mechanisms would act together with the mechanical resistance of the coarctation to ensure an elevated blood pressure above the coarctation and thereby contribute to maintaining an adequate blood pressure below the coarctation. In some patients, as well as in animal models, there is evidence that this chronic disturbance in homeostatic regulation of blood pressure and blood volume may not be fully reversible after surgical correction of the coarctation, and a persistent hypertension develops. In this particular case, the only occasion when normal blood pressures were observed was shortly after surgery, when vascular tone and body fluids would have tended to be low. Even at the three-month follow-up, the pressures were becoming hypertensive.

The acute treatment for congestive heart failure with threatening pulmonary edema must ensure adequate oxygenation, improve the inotropic state of the heart muscle, and relieve the circulatory system of its excess fluid load. Since our theorization would lead us to guess that aldosterone could have been playing a significant role in the salt and hence fluid retention in this patient, a drug to block the action of aldosterone would seem to be a rational selection. Aldosterone acts chiefly on the distal convoluted tubule to increase the reabsorption of Na^+. However, this movement of cation across the renal tubule must be balanced to preserve electrochemical neutrality. Some of this balance is readily achieved by the simultaneous reabsorption of Cl^- or HCO_3^-. Insofar as reabsorbable anions are not available, K^+ will be drawn into the tubular fluid and excreted with the anion that had originally accompanied the sodium.

Partial suppression of this tubular reabsorption of Na^+ will have three secondary consequences. (1) Ample residual Na^+ will remain within the tubular fluid to balance tubular anions and thereby eliminate the electrochemical force for ion exchange and K^+ excretion. (2) In a similar fashion, there will be less cation reabsorption to foster the reabsorption of bicarbonate anion. (3) Reduced reabsorption of bicarbonate will reduce the excretion of hydrogen ion. As a result, antialdosterone drugs characteristically lead to an increase in plasma K^+ and a tendency toward acidosis. Cellular buffering of the retained hydrogen ion leads to further shifts of potassium into the extracellular volume to add to the potassium accumulated by renal retention. Excess K^+ outside cell membranes shifts the cell membrane potential in a depolarizing direction. Persistent hypopolarization of the cells lowers their excitability by depressing the activation of the Na^+ channels that are necessary to generate action potentials. This process will be accentuated by the acidosis that increases the ionization of calcium and further interferes with Na^+ channel activation. This combined ionic disturbance was responsible for the state of muscle weakness observed in this patient.

The immediate problem of hyperkalemia can obviously be corrected by discontinuing the use of this particular diuretic drug. This episode focused attention on the renal function of this patient, however, inasmuch as his susceptibility to the complication of this drug therapy could have related to deficiencies in his renal function. The observed creatinine and BUN levels offer further evidence that the renal component underlying this patient's hypertension warranted further evaluation.

13

A 35-year-old male office clerk was brought to the emergency room because of a severe fainting spell. He was perfectly lucid on arrival at the hospital and had no complaints, but he confessed that he had had several previous fainting spells in recent weeks for no apparent reason. Six months previously, he had a routine physical examination at his place of employment, and he was told that he had high blood pressure and a heart murmur. In the emergency room he was found to have a pulse rate of 46 beats/min and a blood pressure of 160/110, and he was admitted to the hospital for further study.

Examination in the hospital revealed an enlarged heart. The second heart sound was diminished in intensity and a harsh systolic murmur was heard during systole over the apex of the heart. His blood pressure was 158/124 with a pulse rate of 82 beats/min. An electrocardiogram showed a normal sinus rhythm and left-axis deviation. The routine blood chemistries were all within normal limits. On the afternoon of his first hospital day, the patient again suffered a fainting spell. His blood pressure was 155/90 with a pulse rate of 44 beats/min, and he complained of some shortness of breath. Another electrocardiogram was taken. On the basis of these findings, he was scheduled for diagnostic cardiac catheterization.

In the catheterization laboratory, a catheter was passed into the heart under fluoroscopic control and passed out into the lung fields as far as it could be advanced. The pressure recorded from the catheter in this position was essentially free of cardiac pulsations, indicating that the catheter tip was wedged into a small pulmonary artery. The following pressures and blood oxygen determinations were made as the catheter was withdrawn from this position:

Site	Pressure (mm Hg)	Oxygen content (vol%)
Pulmonary wedge	14	19.8
Pulmonary artery	38/18	13.6
Right ventricle	38/2	13.7
Superior vena cava	3	13.6
Brachial artery	165/125	19.6

Oxygen consumption, 240 ml/min; body surface area, 1.64 m^2.

On the basis of these findings and a general clinical evaluation, the patient was scheduled for surgery, and a major surgical procedure was carried out.

He returned from the operating room with a blood pressure of 148/96 and a pulse of 84 beats/min. His recovery was uncomplicated until the fourth day, when he suddenly complained of severe chest pain. He was short of breath, showed marked distention of his neck veins, and had slight cyanosis of his lips and nail beds. Blood pressure was 118/90 with a pulse rate of 112 beats/min. Examination disclosed an accentuated pulmonic second sound, diminished breath sounds, and dullness to percussion over the right lower lobe of the lung. An electrocardiogram revealed a normal sinus rhythm with slight right-axis deviation and S-T changes, interpreted as right heart strain. Appropriate therapy was administered, and the patient's recovery proceeded without further complications.

In a follow-up examination two years later, the patient was free of complaints. With maintenance on antihypertensive therapy, his blood pressure was 134/88 with a pulse of 76 beats/min. His electrocardiogram was evaluated as being within normal limits. The examining physician was well satisfied with the patient's heart sounds.

Questions

What is the physiological basis for a fainting spell? Is this type of problem to be anticipated in a hypertensive patient?

What was the cause of this patient's fainting spells, and how was it proven?

What is the essential mechanism of a cardiac murmur, and what anatomical sites are suspect when the murmur occurs in systole?

What was demonstrated by the cardiac catheterization?

What type of surgical procedure was indicated? Should this procedure cure the patient's fainting spells?

What was the complication that this patient developed on his fourth postoperative day? How would you explain the distention of neck veins and the cyanosis?

Would the physician have anticipated normal heart sounds on his follow-up examination?

Analysis of Case 13

Fainting spells are the result of sudden cerebral ischemia, most commonly associated with episodes of hypotension. They are, therefore, not to be anticipated in the hypertensive patient. Older hypertensives with considerable cerebrovascular disease may complain of dizziness and may experience small strokes associated with brief periods of unconsciousness, but these complications do not fit the picture presented by this patient.

The two instances of fainting in which data were obtained reveal a marked bradycardia following the faint. At other times a normal pulse had been observed. This would identify the fainting spells as Stokes-Adams attacks associated with transient, complete heart blocks. This diagnosis would obviously have been proven by the electrocardiogram taken at the time of his second attack in the hospital.

Heart murmurs are generated by turbulent blood flow that is created when there is a disruption of the normal pattern of streamlined flow through the vascular system. The intense low frequency components of a "harsh" murmur are particularly suggestive of blood vigorously squirting through a narrow opening. For this to occur during systole suggests either a stenosed aortic valve or a leak in the intraventricular septum. Because different observers did not agree as to the proper interpretation of the heart murmur, and because the A-V block identified functional problems close to the intraventricular septum, cardiac catheterization was carried out to eliminate the latter possibility.

The catheterization demonstrated two abnormalities. The pulmonary wedge pressure indicated some increment in back pressure from the left ventricle. In addition, the Fick data for cardiac output are calculated as $240/(19.6 - 13.6) \times 100 = 4000$ ml/min. (This technique was the precursor to the dye method, which is now almost universally employed for cardiac output determinations.) This value represents a cardiac index of 2.4 liters/min/m^2, which is somewhat low for a man of this patient's age. Of more immediate diagnostic importance was the absence of any evidence of an admixture of arterialized blood appearing in the right ventricle, which excludes an intraventricular septal defect as the source of the systolic murmur.

Thus stenosis of the aortic valve is left as the explanation for the observed murmur. To quantitate the degree of this stenosis, a catheter could have been passed retrogradely into the left ventricle to measure the systolic pressure gradient between the left ventricle and the aorta. Retrograde passage of a catheter through a stenosed valve can be difficult, however, and was deemed unnecessary in this case. There was already evidence of functional disturbances, resulting from the high intraventricular pressures required to eject blood past the stenosed valve. Both the systemic hypertension and the flow resistance of the stenosed valve would have required very high ventricular pressures to eject blood during systole. These high intraventricular pressures result from the development of high tensions within the myocardial wall — tensions that would be augmented by the enlarged heart volume. These high intramural tensions act as a compressive force on blood vessels within the wall and could compromise local blood supply. Such a compromise of local blood supply to the A-V node or common bundle of His would explain conduction failure in the A-V nodal system and the Stokes-Adams syncope the patient had experienced. Thus there is clear indication that the valve stenosis warrants treatment.

During the recovery from excision of the stenosed valve and replacement with a prosthetic valve, the patient had an episode on his fourth hospital day that bore no resemblance to his previous problems. On this occasion tachycardia with a narrow pulse pressure was observed. At the same time the accentuated second pulmonic sound, the distended neck veins, and the electrocardiogram gave evidence of pulmonary hypertension. The explanation for this hypertension relates to the lesion that had suddenly appeared in the right lung, indicating pulmonary embolization. The blood-clotting mechanism is a dynamic system, always tending to some microscopic clot formation and lysis of these potential thrombi. However, stagnation of blood interferes with this dynamic equilibrium and favors clotting. After major surgery, patients tend to be immobilized because of the discomfort of movement; and sedation to ease their discomfort obviously increases their immobilization. This immobilization leads to regions of sustained compression of veins by the gravitational pressure from overlying tissues, and venous stagnation occurs behind the compression points. Unless adequate shifting of body weight can be achieved to alter these points of venous obstruction, the prolonged venous stagnation may lead to venous thrombosis and the threat that thrombi will break loose and be carried to the lungs. To protect against venous thrombosis, and more importantly to protect against thrombus formation on the artificial valve, anticoagulants have important therapeutic value in the management of patients with aortic valve replacement. However, anticoagulation had been delayed in this patient to avoid bleeding from his fresh surgical wounds.

While part of the picture of pulmonary embolization must necessarily be related to the mechanical block caused by the embolus, direct mechanical blockage is totally inadequate to explain the phenomenon described in this patient. Since a significant portion of the pulmonary vascular bed is in a state of partial collapse in the resting state, mechanical blockage of one segment should be accommodated by the opening of other partly collapsed regions, with minimal changes in pulmonary artery and right heart pressure. In theory, moreover, mechanical blockage could clearly not account for arterial cyanosis, since blocked segments of the vascular bed could scarcely contribute venous blood to the left atrium. Confirmation of these relationships can be obtained by artificial clamping of a pulmonary artery, which leads to minimal changes in pulmonary pressures and no cyanosis. In the process of pulmonary embolization, there appears to be a release of humoral agents, reflex effects, or both, which produce a transient, generalized constriction of the pulmonary vascular bed and open shunts, bypassing the pulmonary alveoli. It is these reactions that must be chiefly responsible for the pulmonary hypertension and the cyanosis.

The surgical correction of the valvular stenosis and the prosthetic valve replacement, together with the antihypertensive therapy, had restored this patient's heart to good functional performance. With the high intraventricular pressures relieved, no further conduction block was experienced. It should be appreciated that the sounds associated with a prosthetic valve were highly abnormal sounds, but their persistence in a clean, unaltered form offers the physician some reassurance that troubles were not developing with the valve.

14

The 18-year-old son of an immigrant mill worker was admitted to the hospital complaining of shortness of breath. According to his father, the patient had been born with a "heart murmur," and the parents had been advised to restrict his activities. He had developed normally, however, and neither the parents nor the patient had paid much attention to the admonition, although school physicians had refused to permit him to participate in scholastic sports. His childhood had been normal, and his parents were never aware of any abnormal exertional dyspnea or cyanosis; there was no clubbing of his fingers. Five weeks prior to admission, he suddenly became extremely dyspneic while walking home from high school and felt dizzy. After lying down for a few moments his condition improved enough so that he was able to reach home. Subsequent to this episode he exhibited lethargy, great fatigability, and marked exertional dyspnea. His bouts of dyspnea were accompanied by cyanosis of the lips. A few days later, swelling of the ankles was noted and two weeks later some swelling of the abdomen appeared. He was taken to the family physician, who treated him with digitalis without any significant improvement.

Physical examination revealed a rather slight but otherwise normally developed young man. He was notably pale and in severe respiratory distress. His pulse was 92 beats/min and grossly irregular; his blood pressure was 130/60. There was dusky cyanosis of his lips and nail beds, with venous engorgement in his neck. His chest revealed dullness to percussion at both bases with decreased breath sounds and moist rales. There was marked precordial activity with a palpable systolic thrill over the apex and at the base, with cardiac enlargements to the left. A loud, rumbling systolic murmur was heard near the sternum in the left fourth intercostal space, transmitted over the apex and heard clearly in the back. Femoral pulses were pistol-shot in character. The abdomen was distended with moderate ascites; the liver was palpable 3 to 4 fingers below the right costal margin. There was marked scrotal edema and marked pitting edema of the lower extremities from the toes to the knees.

An admission blood sample demonstrated 5.8 m RBC with a hemoglobin of 16 gm/100 ml. Urinary findings were normal. An electrocardiogram showed atrial fibrillation with slurring of the ventricular complexes. The patient was given digitalis, mercurial diuretics, and oxygen. He became progressively more dyspneic and orthopneic and expired 20 hours after admission.

Questions

What is the explanation for the sudden appearance of cyanosis in previously acyanotic heart disease?

What sudden change in functional status could have converted this symptom-free condition into seriously decompensated cardiac failure?

Why was the patient pale at the time of his hospital admission?

What is the relationship between the altered functional status and the evidence of massive fluid retention?

How much of the terminal picture relates to chronic changes that were well advanced before the episode five weeks before death, and how much indicates acute developments that are related to the terminal complications?

Analysis of Case 14

This case is a tragic example of medical neglect of a curable condition — neglect that is partially attributable to the socioeconomic status of the family. Responsibility must also be shared by the family physician, who was apparently treating symptoms without recognizing the necessity of making a definitive diagnosis before it was too late. Physiologically, the case illustrates how a potentially serious problem can be well compensated for over many years until a tip of the balance sends the condition into terminal decompensation.

Pressures in the systemic circulation are substantially greater than are pressures in the pulmonary circuit. Congenital defects in the anatomy of the heart and great vessels will, therefore, be expected to result in "leaks," or shunts, from the systemic circuit to the pulmonary circuit. This will increase pulmonary blood flow but will not produce cyanosis, since all blood reaching the aorta will have passed through the lungs. Congenital cyanotic heart disease demands that there must also be some structural deformity creating abnormal pressure gradients to cause blood to flow from the pulmonary circulation into the systemic circuit. The sudden conversion of previously acyanotic heart disease into cyanotic heart disease, therefore, would indicate that some functional disturbance had occurred to alter pressure gradients and cause a reversal of the shunt flow through the defect.

For 18 years this boy had lived with his defective heart without any functional handicap of sufficient magnitude to be noted by the patient or his parents. The sudden disturbance that precipitated the decompensation of his cardiovascular function appears to have been the onset of the atrial fibrillation that was evidenced by the grossly irregular pulse found at the time of his hospital admission. When atrial fibrillation is first initiated in a heart with a normal conduction system, bombardment of the A-V node with atrial impulses produces an extremely high ventricular rate. This rate is so rapid it permits very abbreviated intervals for filling. The stroke volume, therefore, drops proportionately more than the ventricular rate increases and cardiac output falls. This would account for the attack of dyspnea and dizziness experienced by the patient at the onset of his difficulties. As this intense driving of the A-V node persists, there is some deterioration of its rate of recovery with a prolongation of the nodal refractory period. Thus a "functional" A-V block develops, which reduces the number of impulses that can reach the ventricle. This slowing of the ventricular rate affords more adequate filling intervals and partially restores cardiac output.

Filling intervals remain irregular, nevertheless, and the lack of a synchronized contraction of the atrial muscle further detracts from optimal ventricular filling. The restoration of cardiac output will, therefore, not be complete. It should be noted that on admission to the hospital, this patient was described as being "pale" even though blood studies indicated some compensatory polycythemia developing in response to his cyanosis. This finding indicates cutaneous vasoconstriction as a reflex compensation to both the cardiac output deficit and the arterial hypoxemia.

In uncomplicated atrial fibrillation, the development of a functional block, together with these systemic compensations for the reduced effectiveness of ventricular pumping, restores cardiovascular function to reasonably adequate levels. The problem is to identify why such a drastic deterioration of cardiac function developed in this patient. Obviously, this occurrence must relate to the defect that had been present in his heart since birth. The nature of the defect is suggested by the arterial pressure and the peripheral pulses. The wide pulse pressure indicates a large stroke volume that is being rapidly dissipated by runoff through some low resistance pathway, as evidenced by the low diastolic pressure and the pistol-shot peripheral pulses. The murmur localizes the low resistance pathway to the vicinity of the heart, indicating some type of shunt between the aorta and the pulmonary artery. As will become evident, prior to the onset of the atrial fibrillation, the degree of shunt from the aorta to the pulmonary artery was probably much greater and, therefore, the pulses were considerably more abnormal than at the time of his hospital admission.

Since pressures are normally much higher in the aorta than in the pulmonary artery, we would expect flow through a connecting shunt to be exclusively in the direction of the pulmonary artery, creating a short loop of high blood flow through the lungs, the left atrium, the left ventricle, and out the aorta to the origin of the shunt. Such a chronic overload of the pulmonary bed stimulates hypertrophy of the blood vessels and narrowing of their lumen, with a consequent increase in pulmonary vascular resistance, producing pulmonary hypertension. The stiff vascular bed in the lungs of the patient with pulmonary hypertension will show a much greater hemodynamic response than normal to the elevated left atrial pressure that results from atrial fibrillation, and the pulmonary hypertension will be aggravated. If this pulmonary pressure rises to a level similar to systemic arterial pressures, the left-to-right shunting will be reduced and, at any moment during the phasic cardiac cycle that pulmonary pressure exceeds aortic pressure, the shunt will be reversed to carry blood from the pulmonary artery to the aorta. Atrial fibrillation, therefore, caused some shunt reversal and the appearance of cyanosis, indicating the addition of some pulmonary arterial blood to the aortic outflow.

There are now three factors compromising left ventricular function in this young man: a lifetime of chronic ventricular overload, inefficient filling mechanisms due to the atrial fibrillation, and hypoxemia of the coronary blood due to the shunt reversal. This impaired cardiac function caused some reduction in cardiac output that would lower arterial pressure. This lowering of arterial pressure is minimized by the baroreceptor reflex, which, when receiving less pressure stimulation, imposes less inhibition on the sympathetic vasoconstrictor system and also less inhibition of the neural stimulation of renin release in the kidneys. The tendency for arterial pressure to fall also acts directly on the kidney to stimulate renin release. The released renin acts on a plasma globulin to break off a polypeptide that is converted to angiotensin, which is a potent vasoconstrictor. Angiotensin is also a major stimulus for the release of aldosterone from the adrenal cortex. At the same time, the sustained distention of the fibrillating and, therefore, mechanically ineffective left atrium leads to adaptation of stretch, or "volume," receptors in the atrial wall, whose stimulation also acts reflexly to suppress renin release. Loss of this control would further augment the release of renin, angiotensin, and aldosterone.

Activation of the angiotensin—aldosterone system enhanced the constrictor action of the adrenergic sympathetic system and also led to salt retention by the kidney. Salt retention, in turn, has an osmotic action to retain water; blood volume and interstitial fluid volume, therefore, increased. There appear to be even further mechanisms to reinforce this fluid retention system, including a shift of renal blood flow toward the juxtamedullary nephrons, which are avid salt retainers, the release of salt retaining prostaglandins from the renal medulla, and perhaps other factors. All these factors result in massive fluid retention. It should be recognized that an immediate consequence of fluid retention is increased blood volume and increased venous return to the heart, which should stimulate greater ventricular contraction by increased diastolic filling and the maintenance of a higher arterial blood pressure and blood flow. The failing ventricle has an impaired ability to respond to increased filling, however, and in this patient the increased blood volume would have the effect of still further aggravating the pulmonary hypertension and increasing the shunt flow from the pulmonary artery into the aorta. The fluid retention, therefore, becomes self-defeating in improving the cardiovascular status of the patient and manifests itself by the development of the massive edema exhibited by this young man.

It is important to note that in this patient, as with most patients with congestive heart failure, it is the extreme dyspnea that is most disturbing to the patient. The primary source of this dyspnea is the severe congestion of the pulmonary vascular bed. This engorgement of the pulmonary vasculature makes the lung relatively stiff, increasing the work of breathing and augmenting the tensions developed in the lung structure with inflation. These increased tensions act to increase the effective sensitivity of the various stretch receptors in the lung, and, therefore, the reflex drive from these receptors becomes much stronger and respiratory stimulation results. In this patient, an additional contributor to the respiratory drive would be provided by the chemoreceptors responding to the hypoxemia that resulted from the shunt reversal.

The congenital lesion that would be encountered most frequently in a patient presenting this history would be a persistent patent ductus arteriosus, due to failure to obliterate this fetal circulatory shunt, which normally closes shortly after birth. Autopsy of this case revealed that the defect was a physiologically equivalent but embryologically distinct lesion, caused by a failure of complete fusion of the septum that divides the primitive truncus arteriosus into the pulmonary artery and the aorta. This can be a surgically correctable lesion in a patient who has not been allowed to deteriorate to the terminal phase of congestive heart failure that the atrial fibrillation produced in this patient.

15

A 48-year-old cab driver was admitted to the hospital for evaluation of "heart trouble." For about four years the patient had become progressively more aware of increasing fatigue on exertion. For approximately two years he had noted swelling of his ankles and some abdominal distention. For the past six months he had been treated with digitalis glycosides by his physician with no improvement in his condition. He admitted to some shortness of breath on exertion, but reported that he slept well and never noted shortness of breath at night.

His pulse was 88 beats/min and his blood pressure was 126/84. On examination of his chest, his heart size was judged to be normal, no murmurs were noted, and his lung fields were clear. Abdominal examination revealed an enlarged liver, an enlarged spleen, and appreciable ascitic fluid in the abdominal cavity. His lower legs and ankles were quite swollen. The examiner noted conspicuously swollen jugular veins when the patient was lying on the examining table and requested the patient to sit erect. Marked distention of the jugular veins persisted in the erect position, and careful observation revealed a conspicuous pulsation of the jugular veins. The examiner asked the patient to take a deep breath and noted a marked weakening of the radial pulse with inspiration. Repeating the maneuver with a blood-pressure cuff in place demonstrated a fall in systolic blood pressure of 15 mm Hg during inspiration.

Laboratory findings included normal plasma electrolytes, a plasma albumin of 4.1 gm/100 ml, globulin of 2.6 gm/100 ml, bilirubin of 0.28 gm/100 ml, and alkaline phosphatase of 3.1 units. A bromsulphalein test showed retention of 14% of the dye after 45 minutes (normal is less than 5%). An electrocardiogram showed low-voltage complexes, and the T waves were almost flat.

The patient was referred for cardiac catheterization. With the catheter wedged into a peripheral lung field, a pressure of 24 mm Hg was recorded. When the catheter was drawn back into the pulmonary artery, a pressure of 30/22 was recorded. Pressure in the right ventricle registered 30/4. Pressure in the right atrium was 22 mm Hg. A cardiac output determination by dye injection gave a cardiac index of 2.4 liters/m^2 (normal is 3.0). Repeating the cardiac output determination while the patient did leg exercises achieved a cardiac index of 2.7 liters/m^2. Analysis of the right ventricular pressure waves recorded during catheterization revealed a sharp fall in pressure toward zero at the beginning of diastole, followed by a rapid rebound toward a level of 20 mm Hg for the remainder of diastole, as shown in Figure 15-1.

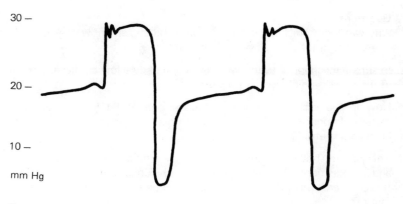

Figure 15-1. Right ventricular pressure curves, recorded at the time of catheterization.

Questions

Why was the bromsulphalein retention test done, and how do you interpret it in the light of other findings?

In simple mechanical terms, what is indicated by a low cardiac index in the face of a high venous pressure?

How do you account for the fact that this patient did not have any major pulmonary congestion or any complaint of orthopnea in spite of his high left atrial pressure?

This patient showed a minimal increase in cardiac output on exertion, and yet exertional dyspnea was not a conspicuous complaint. How does this compare with the usual findings in a patient with congestive heart failure?

What effects does inspiration have on the arterial pressure of a normal individual? How do you explain the "paradoxical" effect observed in this patient?

On the basis of this evidence, what must be the nature of his cardiac problem?

What accounts for the peculiar form of ventricular pressure curve shown by this patient?

Analysis of Case 15

This patient has a physiologically simple problem, but one that presents a complex picture of signs and symptoms. Hence the experienced clinician learns to recognize this disease without necessarily analyzing its detailed manifestations. It is instructive, nevertheless, to dissect the problem and understand why it presents in the fashion that it does.

A patient complaining of fatigue with an enlarged liver and spleen and ascitic fluid immediately suggests cirrhotic liver disease, even though the distended jugular veins would not fit this diagnosis. The bromsulphalein retention test showed only a moderate delay in hepatic excretion of the dye, and other liver function tests were normal. This would be compatible with a secondary effect of chronic passive congestion of the liver rather than primary liver disease.

The distended jugular veins, coupled with the low cardiac output, clearly indicated that there was some type of failure of the cardiac pump to deliver the venous return to the arterial side of the circulation. The high pulmonary wedge pressure indicated a high pressure for filling the left ventricle. This high filling pressure should have distended the ventricle with a large volume of blood during diastole and evoked a vigorous Frank-Starling response of the ventricular muscle during its systolic ejection. The left ventricle of this patient did not show evidence of Frank-Starling compensation, either in the arterial pressure developed or in the cardiac output delivered. Furthermore, with exercise, which increases venous return and should have greatly increased ventricular filling, there was minimal increase in cardiac output. This evidence suggests left ventricular failure.

Two aspects of the patient's history, however, are not in accord with the explanation of left ventricular failure. The patient showed no improvement with digitalis, which is usually a very effective drug to increase the inotropic response of a failing left ventricle. More significantly, the patient complained chiefly of fatigue, which is attributable to his deficient cardiac output response; his examination did not demonstrate pulmonary congestion nor did this patient complain significantly of the severe cardiac dyspnea that pulmonary congestion, secondary to a failing left ventricle, produces (cf. Case 14).

On a simple hemodynamic basis, therefore, the only explanation for left ventricular failure unaccompanied by signs of significant pulmonary congestion must be that there was an equal failure of the right ventricle. The dam resided at the right ventricle and accounted for the distended jugular veins, the enlarged liver and spleen, and the edema in the abdominal cavity and the lower extremities. Yet this cannot be a case of primary right ventricular failure, since the high wedge pressure demands that the process also involved the left ventricle. A symmetrical failure of both left and right ventricles because of myocardial disease would be a highly unreasonable assumption, since the work load of the two muscles is so grossly unsymmetrical. The evidence demands some disturbance that could affect the two chambers equally. This reasoning leads to the pericardium, which surrounds all cardiac chambers symmetrically. If some type of inflammatory process within the pericardial sack had resulted in scarring and constricting of the pericardium, it could throttle the action of both ventricles, prevent their adequate filling, and account for the hemodynamic evidence of pump failure.

The right ventricular pressure contour described for this patient is characteristic of this condition. During systole, myocardial contraction develops appreciable elastic tension in the interfacial planes within the heart wall. When muscle contraction ceases, this elastic tension rebounds to restore the heart to its relaxed state. This elastic rebound, together with the restriction to rapid filling imposed by the constricted pericardium, accounts for the momentary drop of intraventricular pressure toward zero. With the relaxed heart restrained by the constricted pericardium and the large reservoir of blood dammed in the venous reservoir, the pressure quickly rises again to the venous pressure level, and further filling is prevented by the stiff pericardium.

Another characteristic of this disease is the paradoxical pulse. Normally, the increased negativity of the intrathoracic pressure during inspiration sucks more venous blood back to the heart, which results in a phasic rise in cardiac output and arterial pressure. Two possible factors may impede this process. If there is some degree of left ventricular failure but a low pulmonary blood volume, the increased venous return may be stored in the enlarged pulmonary vascular bed of the inflated lung and thereby negate or actually reverse the normal increase in cardiac output observed with inspiration. This mechanism could scarcely be operative in this case since the pulmonary pressures suggest a well-filled pulmonary vascular bed. Another mechanism relates to the fact that the pericardium is attached to the diaphragm, and descent of the diaphragm with inspiration can exert traction on the pericardium. In normal function the pericardium is relatively lax, and this traction has a minimal effect on ventricular function. With a constricted and stiff pericardium, this traction, associated with descent of the diaphragm, would impose an additional pear-shaped constriction on the ventricles and still further impair their capacity for filling. Consequently, in this disease ventricular filling is decreased rather than increased during the latter phase of inspiration.

The surgical procedure for which this patient was referred was resection of his scarred pericardium to permit adequate ventricular filling.

16

A 55-year-old automobile dealer was examined by a physician and found to have a blood pressure of 196/124. The patient admitted to having been told that he was hypertensive a number of years earlier, but he had refused treatment because he felt he was in excellent health. He denied any significant headaches, disturbances in vision, or dizzy spells. An older brother had recently suffered a severe stroke, however, which had made this patient apprehensive about his own condition.

A diagnostic work-up failed to reveal any disease with which the hypertension might have been associated as a secondary manifestation. The diagnosis of essential hypertension was made, and the patient was placed on antihypertensive therapy, including a diuretic and a sympatholytic drug.

For the following three months the patient complained of dizzy spells that persisted in spite of attempts to adjust his therapy. He was, therefore, instructed to discontinue his therapy for a period of two weeks and then admitted to the hospital for a more thorough study. His blood pressure on admission was 184/120. A neurological examination was negative.

He was started on a carefully graded increase in dosage of a sympatholytic drug over a period of several days. By the third day his pressure had been reduced to 152/96 and the patient had no complaints. On the fourth day the patient experienced a number of dizzy spells while attempting to walk around the ward. His blood pressure on that day was recorded as 142/88.

His therapy was discontinued for two days and he was then taken to a laboratory that was equipped with a tilt table. Lying on the horizontal table, the patient's brachial blood pressure was 175/118. The table was then tilted to a 60° feet-down position and his blood pressure was carefully followed. After 20 seconds in this position, he complained of feeling dizzy. At this point his brachial blood pressure was 116/90.

Questions

What is actually being measured in a tilt-table test?

Do you consider the response of this patient to be normal?

Explain the patient's dizzy spells.

What implications do these findings have for the treatment of the patient?

Analysis of Case 16

The tilt-table test is of value for demonstrating the interaction of two physiological mechanisms: (1) the gravitational pooling of blood in the dependent parts in the absence of postural muscle tone to compress the veins and sustain venous return to the heart, and (2) the compensatory response of the baroreceptor mechanism to induce constriction of both the arterial resistance elements and venous capacity vessels to maintain adequate venous return, cardiac output, and arterial pressure in spite of the peripheral pooling of blood. With an adequate tilt stimulus, normal individuals will eventually become dizzy and exhibit syncope because the baroreceptor response cannot sustain an adequate blood pressure to preserve normal cerebral blood flow. Since the cerebral vascular bed exhibits autoregulatory dilation to compensate for modest deficiencies in the arterial pressure head, this dizziness is usually not experienced until mean arterial blood pressure has been reduced to the level of about 60 mm Hg, where the autoregulatory response of the cerebral blood vessels is maximal. At pressures below this level, cerebral blood flow is not maintained, and the dizziness and syncope result from cerebral ischemia.

Increased susceptibility to cerebral ischemia on the tilt table is most commonly observed in one of two conditions: (1) a deficient response of the baroreceptor reflex or (2) a low blood volume that makes the patient particularly vulnerable to venous pooling. Obviously, a patient who had been receiving diuretic therapy might have had a reduced blood volume. With a low blood volume or a deficient response of the baroreceptor constrictor reflex, however, increased susceptibility to the tilt table would be correlated with a more rapid fall in arterial pressure when the table was tilted. The unusual feature of this patient's response was that he experienced his dizziness while his brachial arterial pressure was still reasonably high.

Chronic hypertension leads to both physiological and anatomical changes in the blood vessels to adapt to the high pressure. One prominent change is a thickening of the media of the blood vessels that causes a narrowing of their lumen. This, obviously, would contribute some increase in vascular resistance and, hence, tend to sustain the hypertensive state. While experts disagree as to how important this factor is in the maintenance of hypertension, it should be noted that the adjective "essential " was derived from an older notion that an elevated blood pressure was required to preserve adequate perfusion of blood in hypertensive patients. In any case, there is ample evidence that a number of parameters of cardiovascular regulation tend to become "reset" to maintain their homeostatic regulation at the hypertensive level. This patient appears to have reset his cerebral vascular responses to operate at such an elevated level and, thus, cannot maintain autoregulatory control of cerebral blood flow over the normal pressure range.

A theoretical goal in treating this or other patients with chronic hypertension would be an attempt to judiciously titrate their blood pressure down to lower levels and maintain it at these levels until the regulatory mechanisms reset to normal levels. Clinical experience provides limited evidence that such a reversal can be accomplished in the typical hypertensive patient, suggesting that the "reset" mechanism is not fully reversible. Unless this patient can be adjusted to tolerate pressures in the normotensive range, some degree of hypertension will truly be essential for this patient to preserve adequate cerebral blood flow.

17

This case history was selected to illustrate how badly a medical problem can be managed when treatment is not based on an understanding of its physiology.

One afternoon a physician was called to the home of a wealthy family to examine their 39-year-old unmarried daughter, whom he had previously treated for a variety of minor ailments. Most of the previous complaints had been secondary to problems with emotional instability and alcoholism. She reported that the previous evening she had become nauseated and vomited shortly before retiring and had several further bouts of vomiting during the night. That morning she noted diarrhea; the vomiting attacks continued during the day. The call to her physician had been prompted by an attack of what she referred to as "the shakes," characterized by generalized trembling and a particularly painful stiffness of her hands. She denied any recent intake of alcohol.

The patient appeared very apprehensive and emotionally depressed, but she was lucid and rational and showed no gross neurological abnormalities. Her blood pressure was 90/50. The physician gave her a capsule of a new drug* and called an ambulance for immediate hospitalization.

On admission to the hospital, the patient had developed a confused mental state with a blood pressure of 80/40, a pulse of 122 beats/min, and labored respirations of 36 per minute. Chest examination revealed distant heart sounds, somewhat obscured by vigorous breath sounds that were heard throughout all lung fields. As her examination was being continued, her respiration stopped. Attempts to palpate a peripheral pulse were unsuccessful. The physician diagnosed ventricular fibrillation, administered a large dose of epinephrine intravenously, instituted cardiac massage by chest compression, and called for countershock defibrillation. By the time the defibrillation apparatus arrived and the electrodes were being applied, it was noted that the patient had a strong heart beat with a normal rhythm. Her blood pressure was 115/60 and her respiration was normal, but the patient was comatose. The physician then ordered oxygen administration.

During the night she received 2000 ml of 5% glucose in 1% saline and 3000 ml of 10% glucose in distilled water, all fluids having a sub-

*According to the manufacturer, the drug that was given provided "relief for pain, fever, vomiting, convulsions, cardiac irregularities, and shock."

stantial dose of norepinephrine added. Three times during the night she developed evidence of pulmonary edema, which was treated with limb tourniquets and a variety of cardiac stimulants.

The next morning she continued in coma and had a blood pressure of 110/68. An electrocardiogram showed a sharper R wave and an altered T wave as compared with an earlier electrocardiogram on file, which was interpreted as evidence suggestive of hypokalemia. The physician left orders for continued fluid administration containing norepinephrine, added potassium, and also prescribed cortisone. A urine sample collected in the morning was reported at noon as containing 4+ glucose. Insulin was ordered. Cutting off the norepinephrine led to a decline in the blood pressure, so the continuous intravenous infusion of norepinephrine was resumed.

This therapy was continued for the following two days, with no return of consciousness and an increasingly labored respiration. Increased doses of norepinephrine were required to maintain her blood pressure. On the evening of the third day she developed Cheyne-Stokes respiration and expired shortly after midnight. The physician attributed the death to brain damage, secondary to ventricular fibrillation.

An autopsy revealed early necrotic changes in all the tissues examined, without any other identifiable pathology.

Questions

What are the possible causes of the episode that prompted this patient to call the doctor?

Do you agree that the evidence suggests that this patient experienced ventricular fibrillation during her admission examination?

Analyze the rationale of the therapy given this patient during the first 24 hours.

What accounts for the pathological findings of the autopsy?

Analysis of Case 17

While recognizing that the physician was confronted with a problem that rapidly developed into a medical emergency with little in the way of clues to identify the nature of the patient's difficulty, there was little sense in most of the steps that he took to deal with the situation. Lack of any firm evidence makes any analysis of this case speculative, but it should be recalled that the first suggestion that this patient's problem was anything more than a typical gastrointestinal upset was the episode of "shakes." While this could have been a convulsive seizure associated with delirium tremens or more specific brain pathology, the report sounded suspiciously like a bout of alkalotic tetany, the stiff hands being a manifestation of the carpopedal spasms that accompany alkalotic tetany. Loss of gastric acid from vomiting would have contributed a metabolic component to the alkalosis. A component of nausea is deep breathing that massages the stomach by vigorous descent of the diaphragm; this represents hyperventilation, which would have added a respiratory component to the alkalosis. This hyperventilation could well have been accentuated by the apprehension of this emotionally unstable patient, bringing her to the point of tetany.

Rushing the patient to the hospital by ambulance would scarcely allay her apprehension. The respiration described at the outset of her physical examination, in the absence of any abnormal pulmonary findings, would certainly appear to have been hyperventilation. If the process of her examination had brought some temporary relief of her anxiety, and the emotional drive on her respiratory system dissipated, the acapnia due to hyperventilation could have led to a respiratory arrest until blood CO_2 levels had returned toward normal. Acapnia would also have contributed to her hypotension by reducing vasoconstrictor tone, although probably not to the degree observed. The period of apnea, however, would have exaggerated her hypotension because of the loss of the respiratory pumping action to aid venous return to the heart. Under such circumstances, it can be very difficult to palpate a peripheral pulse. A stethoscope over the heart would have been a much more reliable method for assessing her cardiac status, while the actual diagnosis of ventricular fibrillation cannot be made with any confidence without the electrocardiographic evidence.

The evidence that was available, moreover, strongly argued against ventricular fibrillation. To bedside observers, ventricular fibrillation would be manifest by respiratory gasping due to sudden cerebral ischemia, which is quite the opposite of what was actually observed. The response of the patient to the epinephrine injection also excludes the diagnosis of fibrillation. There is no way in which epinephrine could revert a fibrillating ventricle to a normal rhythm and, in fact, epinephrine, by increasing the excitability of the conductile tissue within the heart, would make the ventricle more susceptible to fibrillation. The prompt blood pressure response to epinephrine would be compatible with the response to be anticipated in recovery from a hyperventilation syncope.

The maintained coma of the patient, nevertheless, certainly excludes a transient syncopal attack as the explanation of this patient's problem. The therapy given did not achieve, and would not be expected to achieve, an alleviation of the coma. Since there was no basis for implicating hypoxia as a cause of this woman's difficulty, there was no indication for administering oxygen. While a prolonged course of severe vomiting, diarrhea, or both can lead to serious dehydration of the patient, the relatively brief history of the gastrointestinal upset does not suggest the need for the massive fluid replacement given. The bouts of pulmonary edema during the first night indicated that too much fluid was being given too rapidly. The physician defended his therapy on grounds that serious complications were developing too rapidly in his patient to wait for laboratory studies, which would involve an appreciable delay in treatment, but therapy lacking either a bedside or a laboratory justification could scarcely be expected to solve the problem.

The same criticism applies to the potassium therapy. Loss of fluids from the gastrointestinal tract involves a disproportionately greater loss of potassium, but for this depletion of potassium from the large reservoir of potassium in body fluids to become significant, a longer course of the illness would usually be required. The electrocardiographic changes that were interpreted as suggesting hypokalemia are not particularly specific and could well have been produced by the norepinephrine that the patient was receiving. If the physician seriously entertained the thought of hypokalemia, moreover, cortisone is a strange drug to administer, since it has some mineralocorticoid action to promote potassium excretion by the kidney.

The glucose in the morning urine was obviously attributable to the large infusions of glucose solutions given during the night. The injection of insulin would also serve to lower extracellular potassium (cf. Case 37) and, therefore, would be a dangerous drug to use in a suspected hypokalemia without more positive evidence of insulin deficiency. In addition, cortisone and insulin have substantially antagonistic actions on carbohydrate metabolism.

The only thing the physician appears to have accomplished was to sustain the blood pressure of the patient by the continuous norepinephrine infusions. This blood pressure maintenance was achieved by constriction of the peripheral vascular bed and, therefore, was at the expense of the blood supply to peripheral tissues. For the four-day period, use of this potent vasoconstrictor maintained an ischemia of the peripheral tissues, which accounted for the findings of generalized tissue necrosis at autopsy. This necrosis so dominated the pathological picture, it masked any suggestion as to what type of catastrophe this poor woman had encountered. Retrospectively, one may be suspicious of the drug the patient was given before being sent to the hospital. This drug was subsequently withdrawn from the market because of its severe depression of CNS function in some patients. Its unfortunate use in this case conformed to the general pattern of this physician's therapy: the treatment of signs and symptoms without taking the appropriate steps to understand what ailed his patient.

Gastrointestinal System and Liver

IV

18

A 12-year-old girl was admitted to the hospital complaining of cramping abdominal pain and loss of appetite. Her mother reported that the problem had persisted intermittently for the past four years. It was first manifest when the girl encountered a school teacher to whom she was antagonistic. Loss of appetite, vague abdominal discomfort, and swings from constipation to diarrhea were associated with each crisis that developed at school. The problem persisted, though to a lesser degree, when the girl was advanced to the next class. Two years ago the child had been given a thorough examination, including radiological studies of the gastrointestinal tract, but all findings were negative. She, therefore, was referred to a clinical psychologist for counseling, but without any apparent benefit.

Her school work was compromised. Although she appeared to be an intellectually capable child, she was usually taken sick on the eve of any major examination. The child had considerable talent as a figure skater, having learned to skate as a small child on a pond located on the family property, but she had been unable to participate competitively in the sport, because she would fall sick shortly before the time of any scheduled competition. In a similar fashion, several family excursions had to be canceled because the girl was taken sick as the time for departure approached.

On examination the patient was observed to be a rather frail child, complaining of vague abdominal discomfort but without any localizing signs. Her white blood count was normal. Stools were loose but of a normal brown color. A repeat of the GI x-ray examination was interpreted as equivocal without any positive diagnosis. It was finally elected to carry out an exploratory laparotomy. A Meckel's diverticulum was found and excised. The girl recovered from the surgery uneventfully.

A follow-up examination a year later found the child in excellent health, having gained 15 pounds. The mother described her daughter as "a totally transformed personality."

Questions
In view of the diagnosis made at the time of surgery, explain the medical history of this child.

Would you classify this child's problem as psychosomatic?

Analysis of Case 18
One of the more dangerous oversimplifications to make in medicine is to draw a sharp dichotomy between organic disease and psychosomatic

illness. This patient presents an illustration of a proved organic lesion that was responsible for an illness, whose exacerbations and remissions were unequivocally of psychosomatic origin.

Function of the gastrointestinal tract is dependent on homeostatic regulation of secretory and propulsive functions to move the chyme along at an appropriate rate for digestion and absorption of foodstuffs and the consolidation of the residue into suitably formed stools. While the primary regulation of this system is derived from intrinsic and hormonal reactions between the gut and its contents, the parasympathetic system can augment while the sympathetic system can suppress this activity. Augmentation of secretory and motility functions leads to diarrhea, while suppression of these functions leads to constipation. In addition, the composition and the concentration of the intestinal bacteria that contribute to the digestive process become adapted to the normal progression of chyme through the tract. Changes in the nature of the chyme presented to these bacteria will alter their metabolism and tend to result in changes in acidity and the formation of gas pockets that distend a segment of the intestine and give rise to intestinal cramps.

Emotional stress acts on this system through either its parasympathetic or its sympathetic innervation. There is evidence that some individuals tend more toward "vagotonia" while others exhibit more "sympathotonia." For the type of nonspecific complaints experienced by this patient, which component was involved in the prime disturbances makes little difference, since a significant swing in either direction will upset the homeostatic balance and disrupt normal function. The commonplace nature of these functional disturbances is illustrated by the gastrointestinal irregularity that most individuals experience in relation to travel, major examinations, or other life experiences that have significant emotional content.

In this patient these trivial disturbances of function were complicated by the existence of a blind pouch extending from the intestine with somewhat limited communication with the intestinal lumen. The latter was demonstrated by the failure of the diverticulum to fill sufficiently with the barium contrast medium to be diagnosed by x-ray. The relatively stagnant material in this pouch was, therefore, a source for potential irritation and local inflammation if proper drainage from this pouch was disturbed. This local irritation created a positive feedback loop to reflexly alter intestinal function and aggravate the local problem. Secondarily, these disturbances suppressed the appetite of the patient, which removed the stimulus for normal intestinal function as well as altered the substrate for the intestinal flora. These circumstances converted a trivial irregularity in intestinal function into a more debilitating and protracted disruption of her normal intestinal physiology.

A sensitive child, particularly at an early adolescent age of emotional insecurity, can readily develop this type of symptom complex, which, by conditioning, tends to be reinforced with each successive emotional crisis. This girl's vulnerability to this type of functional disturbance, however, related more to her intestinal lesion than to her emotional overreactivity.

19

A 48-year-old truck driver visited his physician with complaints of epigastric pain that was most prominent 3 to 4 hours after eating and was
temporarily relieved by an additional snack. Radiological examination
demonstrated a duodenal ulcer. He was given appropriate medication
and dietary advice.

Six months later he was admitted to the hospital with severe epigastric pain, following an episode of hematemesis. On his first hospital
day he passed a black stool. After suitable preparation, he was taken to
surgery for a subtotal gastrectomy, including most of the antrum and
a substantial portion of the fundus. A gastrojejunostomy and a vagotomy were also performed. His postoperative course was uneventful. He
reported complete relief of pain for the first time in two years.

On follow-up examination two months later, he was still free of
pain or other complaints, but a weight loss of 5 pounds was noted. On
follow-up examination four months later, he complained that some
pain had returned and that he had developed severe diarrhea. He had
lost an additional 15 pounds. Stool examination revealed a very loose
light brown stool with floating fragments, although stool analyses demonstrated the presence of proteases and lipases. Gastric analysis revealed a titratable acidity of 29 mEq/liter (normal is 40). Following an
injection of insulin, the basal acid secretion (normal is 2 mEq/hr) rose
from 0.7 mEq/hr to 6.3 mEq/hr. With a histamine test, acid secretion
rose to 7.6 mEq/hr (normal is 25 mEq/hr).

On the basis of this evaluation, he was given appropriate medication and dietary advice and was discharged for follow-up. A year later
he had no significant complaints and had regained his normal weight.

Questions

What are the physiological goals in the medical management of the
ulcer patient?

What was this patient's surgery designed to accomplish?

How do you interpret the disappearance and the reappearance of pain
in this patient?

Explain the significance of the gastric secretory studies. What possible
diagnosis is excluded?

What was this patient's problem at the time of his last examination?

Analysis of Case 19

The original phase of this history is a typical peptic ulcer story, which dictates a trial at conservative medical therapy, based on the known physiology of gastric acid secretion (which is the agent immediately responsible for the ulcerative tissue damage, regardless of the ultimate explanation for the susceptibility of some patients to this disease). The chief components of this medical regimen are (1) use of agents that are capable of neutralizing gastric acidity without being reabsorbed by the intestine and thereby disturb the acid-base balance of the body; (2) avoidance of agents known to reflexly or directly stimulate gastric secretion, including highly seasoned foods and condiments, alcohol, and fried foods; (3) periodic snacks to provide food buffers as the stomach empties; and (4) use of emulsified fats, such as in milk and cream. Fats suppress the gastrin hormone mechanism that stimulates acid secretion, a suppression that is probably mediated by the release of another hormone, pancreozymin-cholecystokinin. Specific drug therapy to suppress vagal tone may also be indicated as well as supportive psychotherapy to help the patient minimize his anxieties.

A number of surgical techniques have been developed for dealing with the peptic ulcer problem, which may or may not include direct excision of the ulcerated mucosa, but which all include steps to suppress acid secretion. The procedure used in this case was to remove the source of the gastrin hormone (antrum) and the innervation (vagus), which act synergistically to stimulate acid secretion, to remove much of the source of the acid secretion itself (fundus) and to provide a bypass to shunt the acid chyme away from the ulcerated area (gastrojejunostomy). An incidental consequence of this surgery is that it effectively eliminates the afferent nerves from the ulcerated region, which affords dramatic relief of ulcer pain and may give the patient and sometimes the physician false security that the problem has been solved.

Return of evidence of peptic ulceration in a patient who has been subjected to surgery in an attempt to minimize sources of physiological stimulation of acid secretion always suggests the possibility that the acid secretion might be the result of uncontrolled secretions from a gastrin-producing tumor (Zollinger-Ellison syndrome). However, this possibility is not supported by the gastric secretory studies in this patient. The data demonstrated that the surgery had been moderately successful in reducing both the basal acid secretion and the maximal secretory capacity demonstrated by the histamine test. The studies also demonstrated that the intense vagal stimulation, evoked by an insulin-induced hypoglycemia, is capable of producing a near maximal stimulation of the remaining acid-secreting glands, which may be judged by comparing the insulin test with the histamine test. This occurrence correlated with the reappearance of pain and indicates significant regeneration of the sectioned nerves. The pain at this time could have originated from residual disease at the site of the original lesion or from ulceration at the margin of the gastrojejunostomy.

The chief problem confronting the patient at this time, however, was weight loss and diarrhea, with evidence of incomplete digestion of intestinal contents in spite of evidence of pancreatic enzyme secretion. This is primarily a problem of intestinal hypermotility, resulting from a loss of the reservoir capacity of the stomach, a somewhat larger ostium in the healed gastrojejunostomy than the surgeon had hoped for, and overconfidence on the part of the patient as to the functional capacities of his gastrointestinal tract. The patient was eating large meals that were being delivered rapidly into the intestine. This intestinal distention is a potent stimulus for peristaltic activity, which causes the chyme to move along the tract at a rate that is too rapid for digestion to be completed and for water and foodstuffs to be absorbed. (This patient did not exhibit signs of the "dumping" syndrome due to the major osmotic shifts in fluid that are sometimes associated with these sudden overloads of the small intestine.)

Therapy for the patient was directed toward the use of an anticholinergic drug to reduce intestinal motility and instruction for him to alter his lifestyle by eating a larger number of smaller meals. Small meals at frequent intervals would compensate for loss of the gastric reservoir and restrict the load presented to the intestine at any one time. With this adjustment in eating habits, adequate bowel function was restored.

20

A 43-year-old male chauffeur was admitted to the hospital complaining of severe abdominal pain, pasty diarrhea, and a yellow coloration of his skin. For about two years he had experienced occasional bouts of upper abdominal pain following a meal. He felt he had obtained some relief from these attacks with soda and peppermint. However, the attacks had been coming with increasing frequency until two days ago, when he had a severe and persistent attack that he finally treated with a large dose of castor oil. The castor oil produced nausea and vomiting for 12 hours, followed by diarrhea with constant, severe abdominal pain.

Examination of the chest was negative; examination of the abdomen was unsatisfactory because of extreme tenderness, but an impression was obtained of a somewhat enlarged liver. X-ray examination of the upper GI tract revealed no definite pathology; what appeared to be an enlarged gallbladder was poorly visualized. No gallstones were seen. The laboratory data obtained are as follows:

	Normal	Observed
Hemoglobin (gm/100 ml)	15	12
Red blood cell count (m)	5.5	4.2
Albumin (gm/100 ml)	4.5	3.2
Globulin (gm/100 ml)	2.5	4.0
Gamma globulin (%)	14	22
Cholesterol, total (mg/100 ml)	200	420
Cholesterol esters (mg/100 ml)	140	184
Bilirubin, unconjugated (mg/100 ml)	0.2	8.2
Bilirubin, conjugated (mg/100 ml)	0.1	8.4
Prothrombin time (sec)	14	23
Alkaline phosphatase (u.)	3	18
Bromsulphalein retention (%)	2	65
Urine: positive for bilirubin and negative for urobilinogen		

Special tests were carried out by passing a drainage tube into the duodenum that was maintained under constant suction. During the hour following an injection of secretin, the following duodenal juice was obtained as compared to normal values given in parentheses: volume, 75 ml (200); H^+, 26 nM (6); bicarbonate, 34 mM (180); trypsin, 6 units (10); lipase, 2 units (5). Four hours later an injection of purified cholecystokinin-pancreozymin was administered, and the following juice was collected during the subsequent hour: volume, 47 ml (80); H^+, 31 nM (10); bicarbonate, 20 mM (75); trypsin, 10.8 units (90); lipase, 3.5 units (15); taurocholate, not detectable (80 mg).

A surgical procedure was carried out, during which the patient required 3 units of blood transfusion. Although the procedure was deemed successful, the patient remained acutely ill with extreme abdominal pain and required heavy sedation. On the second postoperative day his blood pressure was 85/60 and he was given 2 more units of blood. The jaundice had largely cleared by the third postoperative day, with a bilirubin of 1.8 mg/100 ml. On the fourth postoperative day the patient appeared severely depressed and mild jaundice had returned. Bilirubin was now 5.2 mg/100 ml with a conjugated bilirubin of 0.6 mg/100 ml. Hemoglobin was 9 gm/100 ml. He was treated with 2 units of packed red cells and vitamin K. Following this episode he gradually improved.

Questions

What is suggested by the history?

Explain all chemical findings.

What do you think the surgeon expected to find?

What caused the jaundice following his surgery?

Analysis of Case 20

When ingested food reaches the duodenum, the hormone cholecystokinin-pancreozymin is released, which causes contraction of the gallbladder and stimulation of enzyme secretion by the pancreas. With inflammation of either organ or obstruction of the common bile duct, this stimulation will evoke pain. Bile duct obstruction is the most common of these problems. When a patient with a history of this type of pain develops jaundice, obstruction of the bile duct is strongly suggested.

The initial laboratory data demonstrated the abnormalities to be expected with obstructive jaundice. The elevation of bilirubin, cholesterol, and alkaline phosphatase, three of the excretory products in the bile, demonstrated a failure of biliary excretion. Since approximately half of this retained bilirubin had been prepared for excretion by its hepatic conjugation into the glucuronate form, impaired liver function would not appear to be responsible for this excretory failure. Confirmation of this failure of bilirubin excretion in the bile was observed in the urine. The soluble conjugated bilirubin in the plasma was filtered by the kidneys and appeared in the urine, while the urobilinogen that would have been present if bilirubin had been exposed to the action of intestinal bacteria was absent.

The remaining liver function tests demonstrate the moderate degree of dysfunction that would be attributable to the congestive inflammation of the liver, resulting from stasis in the biliary system. A specific response to this inflammation was the elevated gamma globulin. The increased retention of the injected bromsulphalein dye and the deficiency in the esterification of cholesterol indicate impaired hepatic function. The prolongation of the prothrombin time suggests malabsorption of the fat-soluble vitamin K, which is a consequence of lack of bile salts for its reabsorption. The mild anemia could be a manifestation of the contribution of the liver to hematopoiesis.

The failure to demonstrate gallstones prevented a positive diagnosis, and the intensity of the pain that the patient was experiencing at the time of his examination raised the question as to whether the pancreas might also be involved in the disease process. The special tests clearly indicated a deficiency in pancreatic secretory function, even after specific hormone stimulation. This information was of value to the surgeon, since it allowed him to anticipate problems that he might encounter. For example, malignant tumors originating in the head of the pancreas often obstruct bile flow as well as pancreatic secretions.

No malignancy was discovered at surgery; instead, the lower portion of the common bile duct was found to be impacted with sandy grit. The surrounding tissues in the ampulla of Vater at the exit of the bile duct were badly inflamed, with apparent obstruction of the adjacent pancreatic duct and acute inflammation of the pancreas. Presumably the patient's self-treatment with castor oil provided intense stimulation of the intestinal tract, which superimposed an acute inflammatory process on the preexisting bile duct obstruction. The bile duct was cleared and dilated and the gallbladder was removed.

The episode of jaundice that developed on the fourth day following surgery represented a different pattern from the chemistry of his jaundice before surgery. The lack of any substantial accumulation of the posthepatic conjugated form of the bilirubin, together with the appearance of urobilinogen in the urine, attested to the success of the surgery and the continuity of the enterohepatic circulation of the bile pigments. The drop in the patient's hemoglobin, together with the rise in the prehepatic unconjugated bilirubin, suggests that the jaundice was produced by a flood of heme pigments released from the patient's red cells. Confirmation of this source of the bilirubin appeared in the stool, which gave evidence of digested blood, indicating hemorrhage into the upper gastrointestinal tract.

Acute pancreatitis is a severe physiological stress, stimulating an excessive production of glucocorticoids by the adrenal cortex. Major surgery added another stressful stimulus, further augmenting this corticoid secretion. The difficult postoperative course experienced by the patient prolonged the stress. A sustained high titer of glucocorticoids tends to disrupt the integrity of the gastric mucosa, producing so-called stress ulcers. The depressed blood-clotting machinery in this patient would have rendered these stress ulcers more susceptible to bleeding. This episode was, therefore, a transient reaction to the stresses to which the patient had been exposed, whose cause was eliminated as the patient's recovery progressed.

21

A 57-year-old male warehouse employee was admitted to the hospital with the complaint of heartburn, shortness of breath, and abdominal distention. He was an obese, sallow-complexioned individual with a massively distended abdomen, scrotal edema, extensive pitting edema of the lower extremities, and bilateral inguinal hernias. He admitted to heavy consumption of alcohol for many years. The vital signs were normal, with a blood pressure of 135/80. Chest examination was essentially normal. Examination of the abdomen, which was grossly distended with fluid, revealed conspicuous veins radiating from the region of the umbilicus.

His hemoglobin was 12.4 gm/100 ml with a hematocrit of 37%; plasma electrolytes were within normal limits; albumin was 1.7 gm/100 ml; globulin was 3.0 gm/100 ml; serum bilirubin was 1.8 mg/100 ml; prothrombin time was 24 seconds (normal is 14); and alkaline phosphatase was 4 units. A bromsulphalein excretion test demonstrated 23% retention 45 minutes after the injection (normal is less than 5%). Analysis of a 24-hour urine specimen demonstrated the excretion of 1.2 mg of urobilinogen (normal is 1.5) and 13 mM of Na^+; at this time the patient was on a low sodium diet that was estimated to contain about 0.5 mg (22 mM) per day.

Because of the degree of discomfort associated with the abdominal distention, a needle was passed through the abdominal wall and 11.5 liters of a straw-colored fluid withdrawn. Analysis of this fluid showed that it contained sodium in a concentration of 142 mM and albumin in a concentration of 1.2 mg/100 ml. Three hours later the patient was noted to be depressed and apprehensive. His blood pressure was 90/55 with a pulse rate of 110 beats/min. His hematocrit was 44%. The patient was given a unit of plasma.

The next morning the patient reported that he felt better. His blood pressure was normal and the ankle edema had completely disappeared. Examination of the abdomen, however, showed the presence of significant ascites. A plasma albumin determination on that day gave a value of 1.2 gm/100 ml.

Questions

The patient's chief complaint on entering the hospital was heartburn. How do you account for this symptom?

What was the cause of the shortness of breath?

What was the cause of the ankle edema?

From the data given, what do you conclude about the liver function of this patient?

What accounted for the acute hypotensive episode following the abdominal paracentesis?

What do you conclude from the urinary findings?

Analysis of Case 21

This patient had alcoholic cirrhosis of the liver with massive ascites. His heartburn and dyspnea were mechanical consequences of the high intra-abdominal pressure associated with the ascites. Of the several valvelike structures at different levels of the gastrointestinal tract, the anal sphincter is the only truly competent valve. The cardiac sphincter is a communication between the esophagus and the stomach, which impedes reflux of gastric contents in three ways: (1) by its thickened walls and hence narrower lumen, (2) by the generally high tone of its musculature when gastric motility is high, and (3) by a mechanical compression of the lower esophagus because intra-abdominal pressure exceeds intra-thoracic pressure, which adds to the closure by crimping the tube. Prevention of reflux is normally maintained between meals when the tone of the muscle is reduced because gastric motility is reduced even further. When intragastric pressure rises because of a rise in the external pressure (ascites, pregnancy), this balance is lost. Reflux of acid chyme into the esophagus produces heartburn. Also, significant mechanical distortions of the area of the cardiac sphincter (due to increased intra-abdominal pressure, diaphragmatic hernia, and descent of the diaphragm in vomiting) may also cause acid reflux, due to unbending of the normal crimp at the base of the esophagus.

The dyspnea was similarly caused by the increased intra-abdominal pressure, which compromised the descent of the diaphragm. In addition, it should be appreciated that the 11.5 liters of ascitic fluid represented a load of 11.5 kg (24 lb) that this patient was carrying around with him, adding to his physical work load.

For venous blood to return from the lower extremities, venous pressure had to rise above the pressure dam in the abdomen created by the ascites. In addition, the depleted plasma albumin significantly reduced the plasma oncotic pressure that is responsible for retaining fluid in the vascular system at the capillary level. The summation of these two factors was responsible for the scrotal and ankle edema.

Quite obviously there is severe fibrotic scarring in the liver of this patient obstructing portal blood flow, as evidenced by both the ascitic fluid and the distended collateral venous drainage on the abdominal wall. Nevertheless, disturbances in liver function, in general, are rather minimal, except for the increased bromsulphalein retention, which is a sensitive indicator of this type of liver disease. Other hepatic excretory functions (alkaline phosphatase, bile pigments) are close to normal. The borderline anemia and slight prothrombin deficiency can be as readily attributed to the deficient diet of many of these patients as it can to liver disease. This problem should in no sense be interpreted as an indication of healthy liver cells but is, rather, an indication of the vast reserve of liver tissue that is normally present. Thus destruction of a large portion of the functional hepatic tissue must occur before gross signs of liver dysfunction appear.

The hypoalbuminemia could suggest depression of protein synthesis by the liver, but it must be analyzed in the context of the very substantial loss of protein into the ascitic fluid. Unlike most capillary beds, the liver sinusoids and the hepatic capsule show a very significant permeability to albumin. When sinusoidal obstruction dams up the portal stream and portal pressure rises, a "weeping" of the liver surface results with exudation of an albumin-rich filtrate of the blood. This albumin in the peritoneal cavity tends to counteract the plasma oncotic pressure of the blood in the capillaries on the surfaces of the intestine at the same time that the portal hypertension is raising their hydrostatic pressure. There results further filtration of fluid from all these surfaces to augment the accumulation of ascitic fluid.

A crucial factor in limiting this fluid loss into the peritoneal cavity is the rise in hydrostatic pressure in the distended abdomen, which becomes quite tense as the ascitic volume increases. Therefore removal of ascitic fluid to ease the discomfort of the patient removes this check on further fluid accumulation and, in this instance, the ascitic fluid reformed so rapidly as to precipitate the hypotensive episode as fluid was drained from the circulating blood volume. The baroreceptor response to this hypotension would induce peripheral vasoconstriction and lower peripheral capillary blood pressure. This lowered capillary hydrostatic pressure acts to mobilize tissue fluid from the periphery, especially in the lower extremities, where the lowering of the intra-abdominal pressure would also serve to lower capillary venous pressure and encourage fluid resorption.

While sodium excretion in the urine cannot be evaluated accurately without a more rigorous balance study, the low Na^+ excretion in this patient is highly suggestive of Na^+ retention. There is little ground for argument that this patient had very substantial sodium retention in the recent past, since the large quantities of ascitic fluid and edema are extracellular — isotonic with plasma largely on the basis of

NaCl content. Aldosterone probably played a major role in stimulating the renal tubules to conserve sodium. The secretion of aldosterone would have been stimulated as the formation of ascitic fluid and edema reduced the blood volume, with consequent detection by the blood "volume" receptors and alterations in renal dynamics (cf. Case 14). It is also probable that there may have been some deficiency in liver degradation of aldosterone; in advanced states of liver failure, this failure to degrade aldosterone significantly extends its salt-retaining action.

Nervous System

A 29-year-old mechanic was admitted to the hospital complaining of extreme weakness of his arms and legs. He had been well until 10 days before admission, when he suffered a severe GI upset, with nausea and vomiting for a three-day period. This episode was followed by swelling of his legs and progressive muscular weakness. He also noted puffiness about his eyes and face on awakening in the morning that subsided shortly after he got out of bed.

Examination revealed a rather lethargic individual with swelling of the lower back and marked pitting edema of the legs. He was capable of only sluggish and weak movements of his extremities and could not support his weight. His blood pressure was 146/98 with a pulse of 80 beats/min. A neurological examination did not reveal anything more than the profound muscular weakness. After blood and urine samples for laboratory analysis were obtained, the patient was given a trial treatment with an anticholinesterase in the hope of relieving his paralysis. No benefits were obtained. He was, therefore, placed in a respirator for fear that the paralysis might spread to the respiratory muscles.

The following values were reported by the laboratory:

Na	148 mM	Cl	102 mM
K	2.0 mM	HCO_3	29 mM
H^+	34 nM	PCO_2	44 mm Hg
BUN	32 mg/100 ml		
Globulin	3.0 gm/100 ml		
Albumin	1.8 gm/100 ml		

Urine: H^+, 180 nM; Albumin, 4+; numerous RBCs.

Upon receipt of the laboratory findings, a blood sample and a muscle biopsy were obtained and sent to a special laboratory, which reported a plasma potassium of 1.96 mM and an intracellular potassium of 10.73 mM/100 gm of muscle cells. The patient was started on potassium infusions and showed a remarkable recovery from his paralysis over the next 12 hours. He was referred to another service for follow-up.

Questions
What explanation do the laboratory findings provide for the edema?

On this basis, what mechanism would be implicated to explain the patient's hypertension? Could this mechanism have any possible relationship with the patient's electrolyte problem?

What type of electrolyte disturbance is to be anticipated after a severe bout of vomiting?

What type of acid-base disturbance does this patient exhibit, and how does it relate to his potassium level?

What quantitative conclusions can be derived from the muscle biopsy data, which are of crucial importance to this patient's immediate problem?

Analysis of Case 22

The albuminuria and hematuria of this patient suggest an acute glomerulonephritis at the time of his hospital admission; what relationship this could have had to his preceding gastrointestinal upset is problematical. One feature of this type of acute renal disease is the release of renin by the kidneys and activation of the angiotensin—aldosterone system. This contributes to the patient's hypertension, and the aldosterone stimulates sodium conservation and fluid retention by the kidneys. The salt—water retention plus the hypoalbuminemia due to renal losses leads to transudation of fluid out of the capillaries into the interstitium, which accounts for the edema shown by this patient.

Aldosterone-induced sodium retention increases renal loss of potassium. This loss would have been superimposed on electrolyte disturbances created by his gastrointestinal upset. Appreciable amounts of potassium are contained in gastric secretions that are lost from the body in vomiting. At the same time, acid is lost from the body, which produces a metabolic alkalosis. The patient's blood chemistry at the time of hospital admission still reflects this metabolic alkalosis. The delay in correcting this disturbance could relate to the hypokalemia, since a deficiency in potassium cation creates more pressure for the excretion of hydrogen ion by the kidneys (cf. Case 31). The cells respond to a metabolic alkalosis by a buffering mechanism whereby hydrogen ion is released by the cells into the extracellular fluid in exchange for potassium ion, representing still another factor that depresses extracellular potassium.

The muscle biopsy was scarcely necessary to indicate this patient's need for potassium, but it was carried out for academic confirmation of the explanation for the muscular weakness that brought this patient to the hospital. The finding of 10.78 mM of potassium per 100 gm cells corresponds with 107.8 mM/liter of cells. Since muscle cells are 70% water, this represents a concentration of 154 mM of potassium per liter of cell water. The plasma concentration of 1.96 mM, when corrected for the fact that plasma is 94% water, represents a concentration of 2.09 mM/liter of plasma water. The Nernst potential for this concentration ratio is −115 mV. The Nernst potential for potassium does not exactly equal the resting membrane potential of the muscle cells, because a few millivolts of potential (of opposite sign) are contributed by a slight leakage of sodium ions. Also, the ionic concentrations that are external to the cell membrane are not quite the same as the plasma concentrations because of the Donnan effect of the plasma proteins. These small quantitative inaccuracies, however, would not alter the conclusion that the muscle cells of this patient were significantly hyperpolarized. This hyperpolarity would have raised their threshold to stimulation, creating a conduction block adjacent to the end-plates within the muscle. This phenomenon, which probably was similarly serving to depress neural function, affords an explanation for the muscular weakness that was of such ominous proportions in this patient to prompt the use of a respirator.

Once the potassium concentration had been corrected, the patient was referred to the nephrology service for further evaluation of his kidney disease.

23

A 62-year-old real estate broker had an annual physical examination and was found to be in excellent health. His heart rate was 74 beats/min and his blood pressure was 140/85. His lung fields were clear and his heart was of normal size with a normal ECG. Six months later he appeared at his physician's office and reported that he had become dizzy and collapsed the previous day while hurrying across a busy street. The physician examined him and found no abnormality to explain the attack of syncope. He reassured the patient and asked him to return if further difficulties were encountered. After leaving the physician's sixth floor office, the patient collapsed in the elevator as it reached the street level. The patient was returned to the physician's office for further examination. His blood pressure was 145/90, his pulse rate was 82 beats/min, and the ECG was normal. In view of the lack of any definite diagnostic indications, the patient was advised to return home but to keep in close touch with his physician.

During the succeeding six weeks, the patient had similar attacks with increasing frequency. Whenever he tried to walk rapidly, he would start to stagger and often fall. He learned that he could forestall these attacks of dizziness by crouching down and placing his head between his knees. He had experienced several fainting spells during defecation and occasionally collapsed while urinating at a urinal. The problem was intensified in hot weather, making it impractical for him to leave his home on a warm day. He was admitted to the hospital for diagnostic evaluation.

Routine examinations, including a thorough neurological examination, revealed no indication of the cause of his difficulties. A series of special tests were carried out to establish the diagnosis. When placed on a tilt table and tilted to a 45° feet-down position, he fairly promptly developed syncope, with blood pressure falling from 134/80 to 48/30 and pulse rate increasing from 68 to 72 beats/min. An exercise step test was stopped when the patient complained of dizziness; his blood pressure had dropped from 124/72 to 86/42 while his pulse had increased from 71 to 73 beats/min. When asked to blow into a tube connected to a mercury manometer to try to maintain the mercury at the 40 mm Hg level, he again exhibited syncope after a number of seconds, his blood pressure falling from 128/75 to 62/40 and his pulse rate increasing from 70 to 74 beats/min. The patient was given an injection of a small dose of norepinephrine that was calculated to produce a minimal rise in blood pressure; his systolic pressure was observed to rise by 45 mm Hg. The patient was placed under an electric blanket for 30 minutes, at the end of which he complained of feeling excessively warm. His blood pressure was 95/60 at this point, and examination showed a warm, hyperemic but dry skin with no suggestion of any perspiration.

Questions

On reviewing this picture, what abnormality was clearly evident at the time of the syncopal attack on his original visit to his physician?

Most patients will eventually develop syncope on a tilt table; what is uniquely different about the response of this patient?

What caused the hypotension with exercise?

How was the Valsalva test related to his symptoms?

What did the norepinephrine test prove?

Why did hot weather aggravate this patient's problems?

Analysis of Case 23

A remarkable feature of this patient's disease was the observation that at no time were there any appreciable changes in pulse rate. A patient who is rushed to a physician's office after collapsing in an elevator would be expected to exhibit a much more marked tachycardia due to apprehension and excitement. This deficiency persisted in all the laboratory tests, where large changes in heart rate should have been observed. The patient had clearly lost his neural mechanisms for controlling heart rate. It is interesting to note that his functionally denervated heart exhibited a pulse rate within the normal range rather than in the range of the moderate tachycardia that is characteristic of the acutely denervated heart.

A comparable deficiency is evident in the patient's vasomotor control. The tilt-table test quickly induced syncope with a precipitous drop in mean blood pressure and an extreme narrowing of his pulse pressure. Thus there was minimal vasomotor compensation for the gravitational shift of the blood volume. Without this compensation, venous return and cardiac output could not be maintained at an adequate level to preserve a sufficient pressure head in the arterial reservoir to sustain cerebral circulation.

The cardiovascular response to exercise involves the interaction between two major responses: (1) generalized neurogenic constriction of the vascular bed and (2) counteraction of this vasoconstriction by profound vasodilation in the active muscle beds, due to local release of metabolites. This routes a very large flow of blood to the exercising muscles with a correspondingly large venous return and large cardiac output. This response is normally reflected in the arterial blood pressure by a substantial rise in systolic pressure and a very wide pulse pressure, the diastolic pressure falling to near the resting range as blood runs off rapidly to the dilated muscle bed. This patient shows an absence of the neurogenic component in both the minimal cardiac acceleration and the fall in diastolic pressure. This low diastolic pressure was due to lack of the neurogenic vasoconstriction of inactive beds to compensate for the metabolite vasodilation of the active muscle bed.

Creation of a positive intrathoracic pressure by forcible expiration with an obstructed airway (Valsalva maneuver) produces a transient rise in arterial pressure as blood is squeezed out of the chest, followed by a sustained decrease in cardiac output, because the positive intrathoracic pressure creates a dam at the entrance of the great veins into the chest to impede venous return. The degree to which arterial pressure falls during this latter phase of the maneuver is a function of how well the baroreceptor reflexes can compensate for this hypotensive tendency. This patient showed an absence of this compensatory vasoconstrictor response. This same mechanism was responsible for his syncopal attacks with defecation and urination. Performance of the Valsalva maneuver by "straining" to compress the abdominal contents is a common voluntary action to provide mechanical assistance for defecation and micturition. This maneuver is performed more often by patients with this syndrome because their disease may also involve deficiencies in the neurogenic control of intestines and bladder.

The patient's exaggerated response to the norepinephrine injection excludes the end organs in the effectors as being the site of this man's deficiency; rather, the patient showed evidence of denervation hypersensitivity. The temptation to attribute his problem to a specific deficiency of the baroreceptor reflex system, which could account for most of his symptoms, was excluded by the failure of any sweating in response to the electric blanket. This identified the problem as a more generalized deficiency in autonomic control, which is due to a lack of efferent outflow.

The thermal response of the blood vessels to an increase in temperature is governed by two mechanisms: (1) a neural vasodilation that is controlled by the hypothalamic temperature regulatory centers and (2) a direct action of the heat to relax the muscles in the wall of the cutaneous blood vessels. The direct local effect of heat would persist in this patient and constitute a lowering of vascular resistance. Normally, a compensatory constriction of blood vessels in deeper structures would be induced, via the baroreceptors, to maintain the arterial pressure head. However, the drop in this patient's blood pressure while he was under the electric blanket resulted from a failure of this compensatory vasoconstriction. A warm environment, therefore, greatly aggravated his other problems of maintaining an adequate arterial blood pressure.

24

A woman in her 30s was suspected of drunkenness by the employees of a department store because she exhibited slurred speech and irrational behavior. She was brought to the emergency room of the hospital by a police ambulance.

A history of the patient was impossible to obtain because she seemed to be disoriented and delirious and was apparently having visual hallucinations. In addition, her tongue and mouth were very dry, making speech difficult. Her eyes had a glazed appearance with widely dilated pupils. Her pulse rate was 132 beats/min, blood pressure was 123/72, respiration was 26 per minute, and temperature was 38.9°C. Her skin was pale and dry in spite of the fact that it was a warm day. The peripheral veins were difficult to identify and appeared to be collapsed. Her lungs were clear, and minimal bowel sounds were heard on abdominal examination. The pupils did not constrict to light, and tendon jerks were interpreted as hyperactive.

A test injection of methacholine was administered; the patient's pulse rate remained in the range of 130 beats/min and no other changes were observed.

Appropriate therapy was instituted.

Questions
What system is identified as malfunctioning by the described signs and symptoms?

How do you account for the patient's blood pressure values in view of the tachycardia she exhibited?

How would a normal individual react to the injection of an effective dose of methacholine?

Analysis of Case 24
This patient demonstrates loss of cholinergic postganglionic autonomic function. The normal human pulse rate is the result of a significant level of sustained vagal tone; removal of this vagal tone releases an intrinsic sinus rhythm that is typical of that shown by this patient. The parasympathetic system controls secretions of the mouth and lacrimal glands and is also responsible for the constrictor tone of the pupil, including reflex constriction of the pupil when light strikes the retina. While motility of the gastrointestinal tract originates in intrinsic mechanisms, significant reinforcement of these local mechanisms is provided

by parasympathetic innervation. The fever exhibited by the patient was probably a consequence of the inability of her temperature-control system to induce perspiration to maintain heat loss on a warm day. The eccrine sweat glands are controlled by cholinergic innervation even though this innervation is derived anatomically from sympathetic nerves. The respiratory stimulation and the behavioral abnormalities of the patient were presumably attributable to a block of cholinergic transmission systems within the brain.

This patient had a normal pulse pressure in spite of a substantial tachycardia, which signifies an increased cardiac output, yet her blood pressure was normal. A high cardiac output with a normal blood pressure dictates that her peripheral vascular resistance was low. The fall in peripheral resistance cannot be attributed directly to a block of cholinergic transmission, since the few components of the parasympathetic system that act on blood vessels are vasodilators. Had there been no decrease in peripheral vascular resistance, however, the increase in cardiac output that is induced by the tachycardia would have increased arterial blood pressure. The fall in peripheral vascular resistance, therefore, was evidence of the homeostatic regulation of the baroreceptor reflex system to maintain arterial blood pressure at a normal level in spite of this increased cardiac output.

Injection of methacholine into a normal subject should evoke flushing of the face, pupillary constriction, sweating, lacrimation, salivation, and a transient slowing of the heart. Failure to observe any of these symptoms provided confirmation that the patient had a blockage of her peripheral cholinergic autonomic receptors.

Following recovery from this episode, it was ascertained that three days previously the patient had been given a prescription for an atropine-like drug for the control of intestinal cramps and diarrhea. Apparently she failed to understand the dosage instructions and had given herself an overdose of the drug.

Kidney

During a routine medical examination, a 28-year-old factory worker was found to have significant glucose in his urine. Otherwise, his examination was entirely negative and the patient claimed that he considered himself in excellent health. He was asked to report for a reexamination on a subsequent morning without having breakfast, at which time his urine was found to be positive for glucose again. A blood sample contained 74 mg/100 ml of glucose. The patient was referred to a medical center for study.

Renal clearance studies were carried out with the patient catheterized and constant intravenous infusions were maintained, initially containing creatinine in saline and then glucose plus creatinine in saline. The data obtained for two 15-minute collection periods are as follows:

	Initial	With glucose
Plasma creatinine (mg/100 ml)	8.6	8.8
Plasma glucose (mg/100 ml)	92	319
Urine creatinine (mg/100 ml)	516	298
Urine glucose (mg/100 ml)	124	1200
Urine volume (ml)	33	60

Questions

What is the most common explanation for glucosuria observed at the time of a routine examination?

Why was this patient not more thoroughly investigated for diabetes mellitus?

What information can be obtained from the clearance data, and what does it indicate as the fundamental defect in the kidneys of this patient?

Analysis of Case 25

The appearance of glucose in the urine is evidence that the load of glucose presented to the kidney (glomerular filtration times the plasma concentration of glucose) exceeded the tubular capacity to reabsorb it. Glucosuria is, therefore, to be anticipated whenever a quantity of readily absorbable glucose is ingested, as in a candy bar, since the blood glucose may exceed the threshold for renal excretion during the absorption peak before the hormonal controls can restore a normal blood-sugar level. To exclude this commonplace phenomenon of an alimentary glucosuria, it is essential to check the urine at a time when there has been no recent ingestion of sugar, as was done with this patient.

The glucosuria of diabetes mellitus is primarily the result of a failure of the hormonal controls to keep the blood-sugar level within the normal range below the renal threshold. It is only at very late stages of the chronic disease that enough secondary damage may have been done to the kidneys to lead to spilling of glucose at normal blood-sugar levels. The finding of glucosuria in an individual with a plasma level of 74 mg/100 ml and without a diabetic history effectively excludes diabetes mellitus.

The creatinine clearance values in this patient are calculated as 133 ml/min in the initial determination and 135 ml/min for the period of glucose infusion, demonstrating normal glomerular filtration. If we use the creatinine clearance as an approximate measure of glomerular filtration, we find that, during the initial period, the filtration of 133 ml/min of plasma containing 92 mg/100 ml of glucose would have presented the renal tubules with a load of 121.4 mg/min of glucose. During the 15-minute clearance period, the kidneys were excreting 2.7 mg/min of glucose into the urine. The renal tubules were, therefore, reabsorbing 118.7 mg/min of glucose, or 98% of the filtered load. Normal kidneys, however, should show complete reabsorption of the glucose at this initial plasma concentration, causing it to be undetectable in the urine by routine analytical methods.

During the second clearance period, 135 ml/min of plasma containing 319 mg/100 ml of glucose was being filtered, representing a glucose load of 433 mg/min. Under these circumstances the patient's kidneys were excreting glucose at the rate of 48 mg/min. The difference of 385 mg/min was being reabsorbed by the renal tubules, a value approximating the normal tubular maximal reabsorptive capacity (T_m) for glucose.

The data, therefore, demonstrate that the patient had a normal tubular capacity for glucose reabsorption, but that this active transport system required substantial glucose concentrations to achieve saturation, resulting in a splay of the reabsorption curve at lower plasma levels. The major cause of this splay relates to the avidity of the carrier system that is responsible for the active transport of glucose across the brush border of the renal tubular epithelium. The normal carrier has an extremely high avidity for glucose, being capable of combining with virtually all the filtered glucose up to the point of complete saturation of the carrier. In this patient there was an abnormality in the structure of the carrier molecules, that required a somewhat higher concentration of glucose to drive the reaction of the glucose with the carrier, which resulted in some glucose failing to complex with the carrier even though the carrier was well below saturation. This free glucose in the tubular fluid cannot be reabsorbed against the higher concentration in the blood without active transport and, therefore, passes into the urine to cause "renal" diabetes.

26

A 42-year-old housewife went to her physician complaining of back pain, urinary frequency, and burning on urination. Her urine was cloudy and tested positive for glucose. Microscopic examination of the urinary sediment revealed a large amount of debris containing numerous white blood cells. The diagnosis of cystitis was made and she was treated with urinary tract antibiotics. In a follow-up examination one month later, she reported that she was completely free of urinary tract symptoms but complained of continuing back pain, lethargy, and some weight loss in spite of a good appetite. Her urine was clear and contained minimal sediment but gave weakly positive tests for albumin and glucose. Hospitalization was advised for a diagnostic workup.

On examination in the hospital, she had a pulse rate of 76 beats/min, a blood pressure of 126/84, and respiration of 18 per minute. Her physical examination was negative except for tenderness in the lower back that radiated into her thighs. Blood chemistry determinations were as follows:

Na	142 mM	Cl	122 mM
K	3.8 mM	HCO_3	14 mM
Ca	3.4 mM	PO_4	1.4 mg/100 ml
H^+	57 nM	Pco_2	33 mm Hg
BUN	16 mg/100 ml	Glucose	98 mg/100 ml
Creatinine	1.4 mg/100 ml	Hemoglobin	11.8 gm/100 ml

Because glucose was found in the urine, a glucose tolerance test was carried out, and plasma glucose levels fell at a normal rate following the test dose of glucose.

The patient's urine had a specific gravity of 1.021, 1+ albumin, 2+ glucose, and contained H^+ in a concentration of 110 nM/liter. A qualitative chromatographic test for amino acids gave intense staining for a large number of amino acids. A 24-hour urine collection demonstrated the excretion of 117 gm of creatinine with a plasma concentration of 1.1 mg/100 ml of creatinine, 1.8 gm of phosphate (normal is 1), and 1.8 gm of alpha amino acids (normal is 0.2).

X-ray of the spine demonstrated rarefication of the vertebral bodies, with partial collapse of one vertebra. Long-bone x-rays demonstrated diffuse patchy rarefaction.

The patient was discharged with appropriate instructions.

Questions

What renal functions are essentially normal?

What anatomical site appears to be the focus of this disease?

No tests for ketoacids were reported; should these determinations have been made?

How do you explain the rarefaction of the skeleton?

Analysis of Case 26

Patients excreting urines with significant nutrient content obviously provide bacteria with a good substrate and, hence, have a higher susceptibility to cystitis. This ailment is too commonplace to be assigned much significance, however, except for the fact that it may bring more important renal problems to the attention of the physician.

This woman's blood chemistry demonstrated no retention of creatinine or urea; quantitatively, her creatinine clearance is calculated as 155 liters/day, or 107 ml/min, which is within the normal range for a woman. She, therefore, had no deficiency in glomerular filtration. She was demonstrating a capacity to concentrate and to acidify her urine, and her low potassium value certainly establishes the fact that she was having no difficulty in excreting the excess potassium in her diet. These three functions are all associated with the distal renal tubules, whose physiology appears to be normal.

The data demonstrate gross disturbances, however, in the proximal tubular function of this patient's kidneys. Important proximal tubular functions are the reabsorption of glucose, amino acids, albumin, potassium, bicarbonate, and phosphate and all these functions were depressed. While the deficiency in glucose reabsorption was least, this transport system has a high factor of safety in its functional capacity and, therefore, is one of the last mechanisms to exhibit gross dysfunction in nonspecific renal disease.

The deficiency in bicarbonate reabsorption can be inferred from the metabolic acidosis that is evident in the blood chemistry. Since this was a hyperchloremic acidosis without a significant anion gap to identify an endogenous source of metabolic acid, renal acidosis is suggested. As pointed out previously, the particular urine sample analyzed showed some acidification, but this degree of acidification was grossly deficient for a patient manifesting an appreciable metabolic acidosis. We may assume that the distal tubular acidification mechanism was still functional, but its capacities were overtaxed by the necessity to titrate significant amounts of bicarbonate that had escaped reabsorption in the proximal tubule. Acidification of a bicarbonate solution generates CO_2, which readily diffuses into the surrounding tissue and leaves no anion behind to carry out NH_4^+ cation; it is this NH_4^+ that constitutes the quantitatively important renal mechanism for excreting significant equivalents of hydrogen ion to correct a metabolic acidosis. As a further complication of this acidification process, the probability of deficient absorption of potassium ion in the proximal tubule would bring fluid to the distal tubules that contained significant potassium cation, which would reduce the electrochemical forces pulling H^+ from the distal tubular cells into the tubular lumen.

The most serious problem this patient confronted, however, related to the demineralization of her bones, which had already led to partial vertebral collapse, producing pain because of dorsal-root compression. This problem relates primarily to her hypophosphatemia, which is due to a failure of proximal tubular reabsorption of phosphate. Although bone mineralization and demineralization are supported by enzymatic machinery, this process is dependent on the solubility properties of calcium phosphate salts. The presence of normal phosphate levels in the plasma and interstitial fluids drives the equilibrium toward precipitation of the insoluble salts; conversely, a fall in plasma phosphate sends the equilibrium in the opposite direction and allows calcium to come out of the bone. Failure of the proximal tubular reabsorption of phosphate, therefore, led to appreciable loss of phosphate in the urine of this patient, as was demonstrated. This led to hypophosphatemia and, consequently, hypercalcemia, as shown in the blood chemistry. This chronically high level of calcium in the plasma would lead to a loss of calcium from the body via the kidneys and the intestinal tract, and, therefore, there would be a steady drain on skeletal calcium. The acidosis resulting from the renal dysfunction would aggravate this loss because calcium solubility increases as the hydrogen-ion concentration increases. The treatment of this patient must obviously be designed to minimize this consequence of her proximal renal tubular disease.

27

A 39-year-old man was prepared for surgery to remove a pulmonary lesion. Shortly before the time scheduled for his operation, the victim of a serious automobile accident arrived at the hospital with massive injuries. It was decided that the operating room originally scheduled for the pulmonary case was more appropriate for the accident case because of its greater supply of accessory equipment, and at the last minute the pulmonary case was reassigned to another operating room. His thoracotomy proceeded smoothly and his lesion was successfully excised. During the procedure he was given 2 pints of blood to compensate for the inevitable blood loss from this type of surgery. Shortly after the patient had been transferred to the recovery room, he passed 400 ml of bloody urine. A frantic check on procedures revealed that in the reassignment of operating rooms there had been a failure to reassign the bloods, and the patient received blood that had been crossmatched for the accident victim.

The patient was carefully followed for the succeeding two weeks with no specific therapy other than regulation of fluid intake; the data recorded are presented in the table.

Questions

What is the fate of incompatible blood that is transfused to a patient?

Why does this affect kidney function?

What was the rationale of the therapy on day 2?

Why was this therapy abandoned in favor of an alternative therapy after this time?

Compare the glomerular filtration rate on the fifth and the fifteenth days. What conclusions may be drawn from these figures?

Why did an acidosis develop on the fifth day? Why was it not actively treated?

What renal functions can you identify as having been disrupted by this accident?

What renal function had still not shown signs of recovery by the fifteenth day?

	Preop	Day 2	Day 5	Day 10	Day 15
Fluid intake (ml)		4830	1380	1900	2020
Urine volume (ml)		220	525	1350	1950
Urine specific gravity	1.021	1.035	1.028	1.010	1.011
BUN (mg/100 ml)	14	57	123	87	26
Creatinine (mg/100 ml)		6.2	8.3	7.5	2.2
Creatinine/24 hr (gm)			1.2		2.8
Urinary glucose	0	Trace	0	0	0
Urinary protein	0	4+	3+	1+	0
Plasma albumin (gm/100 ml)		2.4	1.9	2.1	2.9
Plasma Na (mM)	143		141		152
Plasma K (mM)	4.2		5.9	5.5	4.5
Plasma HCO_3 (mM)	23.6	16.4	13.2	17.6	23.0
Plasma Ca (mM)			1.6	1.7	2.0
Plasma PO_4 (mg/100 ml)		8.5	9.7	9.0	5.3
PSP excretion					
Percent at 15 min			5		34*
Percent at 2 hr			24		68**

*Normal is 40
**Normal is 75

Analysis of Case 27

This type of accident from the transfusion of incompatible blood should never occur in a hospital but, undoubtedly, will continue to happen from time to time as long as human fallibility persists. The massive agglutination and destruction of the incompatible red blood cells released large amounts of free hemoglobin into the blood. Hemoglobin is within the upper limit of molecular sizes that are filterable by the renal glomeruli. While the renal tubules are capable of reabsorbing the small traces of protein that normally leak through the glomeruli, they would be quite incapable of dealing with this load of protein. The hemoglobin would, therefore, be retained within the tubules and some of it would pass into the urine. Recalling that the renal filtrate is concentrated some hundredfold as it passes along the tubular system, however, filtration of significant amounts of hemoglobin results in this nonabsorbable protein becoming very concentrated to form a proteinaceous gel that mechanically blocks the tubules. This blockage disrupts urine formation and tubular function and leads to secondary inflammatory and degenerative changes in the tubular epithelium. The renal tubules possess remarkable abilities to discard these tubular plugs as urinary casts and regenerate functional tubular epithelium, providing the patient is carefully managed to minimize the consequences of the temporary renal failure until renal function is restored.

Following discovery of the problem, the immediate therapy was directed toward inducing a copious urine flow to minimize the concentrating functions of the kidney and wash out as much of the filtered hemoglobin as possible. By the end of the second day, it became evident that this rationale was not working, since only a small fraction of the 4830 ml of fluid administered was appearing in the urine. For a patient on bed rest, one may estimate some 800 ml of insensible water loss and 200 ml of water gained from oxidative metabolism, for a net loss of 600 ml in addition to the urine volume. When this amount is added to the 220 ml of urine volume, there remained a net fluid gain by the patient of 4010 ml on the second day. Were this therapy to be continued, the patient would have become massively overloaded with fluid and would eventually drown from pulmonary edema. Subsequent therapy was, therefore, carefully adjusted in an attempt to maintain the patient in water balance.

This modified management is evident in the data for days 5 and 10, which show a net fluid gain of only 155 ml on day 5 and a net fluid loss of 50 ml on day 10. By day 15 the patient was taking all fluids orally ad lib., and it is interesting to note that he was exhibiting a mild diuresis, for a net fluid loss of 430 ml.

Day 5 obviously represented the height of the patient's uremia, with urea rising to 123 mg/100 ml and creatinine to 8.3 mg/100 ml due to renal retention of these excretory products. This increase relates directly to the creatinine clearance of $1200/8.3 \times 0.1 = 14.5$ liters/day, or approximately 10 ml/min. Another excretory product observed to be rising is the PO_4 ion, a product of protein and nucleic-acid degradation. The fall in calcium is an inverse reflection of this rise in phosphate, since calcium and phosphate in excess of their solubility products will precipitate out of the circulation. The rise in potassium is also an indication of failure to excrete dietary intake, complicated in this case by the release of potassium by tissue breakdown following major surgery and the increased extracellular potassium that is secondary to the acidosis. Secretory function of the tubules was also markedly depressed, as shown by the PSP excretion test. The few tubules that were still functioning at this time were still capable of reabsorbing glucose. The urinary protein excretion must have included significant albumin leaking through damaged glomeruli, as evidenced by the fall in plasma protein.

By day 5 a significant metabolic acidosis had developed because of failure to excrete acid products of metabolism. This acidosis served one protective purpose, however, by increasing the ionization of the limited amount of calcium in the extracellular fluid. Vigorous treatment of this acidosis would have reduced the ionization of this limited amount of calcium and could have precipitated tetany.

The data for day 10 demonstrate that the patient had clearly turned the corner with a progressive return of renal function that was continuing well at day 15. Glomerular filtration had risen to 127 liters/day, or 88 ml/min, by the fifteenth day. The one renal function that, at this point, had not yet shown signs of recovery was the ability of the kidney to concentrate the urine. Specific gravities on days 2 and 5 are relatively meaningless on the basis of the data given, since any urine that contains significant amounts of protein will exhibit an elevated specific gravity. On both days 10 and 15, however, the specific gravity was essentially at the isotonic level. Yet on day 15 the patient exhibited a modest rise in plasma Na level. Since this cation must be balanced with equivalent anions, the patient must necessarily have been somewhat hyperosmolar on this date. With this evidence of hyperosmolarity, the patient should have been retaining fluid and excreting a concentrated urine. The concentration gradient in the medullary loops of Henle that is necessary to excrete a concentrated urine, however, is one of the last functions to be restored after this type of insult to renal function. Until this renal function was restored, the patient had to rely on his thirst mechanism to maintain his osmolarity.

28

A 34-year-old man came to the clinic complaining of burning pain in the urethra during and for a short time following urination. He was asked for a urine sample that was found to be loaded with pus and some red blood cells. On physical examination he was seen to have a surgical scar over his lower spine. The patient explained that six months earlier he had stumbled while leaving the stands in a baseball park and landed on the base of his spine. He had experienced severe shooting pains radiating into his lower legs, which had been relieved by surgery. A check of his hospital records revealed that he had been treated with a decompression laminectomy to relieve pressure on the base of his spine. The patient was given a prescription for urinary tract antibiotics and asked to report back for follow-up.

In a visit to the clinic two weeks later, he reported a complete disappearance of all symptoms. He was asked to empty his bladder, and he delivered about 100 ml of clear urine. Microscopic examination of the urinary sediment showed nothing of significance. The physician then passed a catheter that, on entering the bladder, released about 700 ml of urine. After the physician was satisfied that the bladder was emptied, he arranged to record a cystometrogram to determine the pressures as fluid was slowly introduced into the bladder, as shown in Figure 28-1. With the introduction of the first 700 ml of fluid, there was a minimal rise in pressure. By the time 800 ml had been introduced, there was a modest rise in pressure and a slight leakage of fluid around the catheter. The patient was asked to try to stop this leakage; the leakage promptly stopped and bladder pressure fell. Pressure rose more steeply above a volume of 1000 ml. At this pressure the patient could voluntarily stop leakage around the catheter, but his efforts did nothing to lower bladder pressure. The patient denied any discomfort at this time.

On careful questioning, the patient admitted that since his accident he had experienced different sensations related to urination. He no longer felt an urge to pass his urine because of a feeling of fullness in his bladder. He became aware of the need to urinate by the feeling of urine entering his urethra, and frequently there was some small dribbling of urine. He also complained that his urine came much more slowly and that he had to try much harder to urinate. He demonstrated the nature of the latter by a strong contraction of his abdominal muscles.

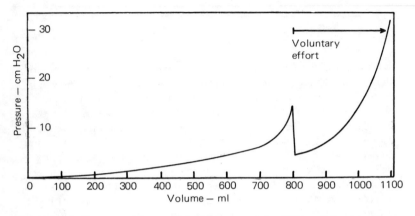

Figure 28-1. Pressures observed during progressive filling of the bladder; above the 800 ml volume, the patient was asked to attempt to prevent any leakage of fluid around the catheter.

Questions

What type of lesion is responsible for this patient's problem?

What does this patient's ability to inhibit the first rise in bladder pressure prove?

Why could not the patient inhibit the final rise in pressure?

Why does the patient have to "try hard" to urinate?

Why is this patient's bladder so large?

Analysis of Case 28

As a residual to his spinal injury, this patient had lost the sensory innervation from his bladder. However, it appears that the motor innervation to his bladder was still intact.

Bladder tone is the result of two factors: (1) the intrinsic response of the smooth muscle to contract when stretched, and (2) a reflex response from receptors within the bladder wall that are sensitive to wall tension and reflexly stimulate the contraction of the bladder muscle. Micturition is controlled with the aid of central connections from these tension receptors. As the bladder becomes distended and tension in the wall rises, sensory information to the cerebral cortex gives rise to descending inhibitory impulses to depress activity in the motor outflow to the bladder and to inhibit the local reflex, with a consequent fall in tension in the bladder wall. As this process continues and distention of the bladder becomes greater, inhibition of the motor supply to the bladder does not totally relieve wall tension, and a feeling of bladder fullness is sensed by the individual. This prompts him to seek an appropriate opportunity to empty his bladder.

With loss of the sensory limb of this spinal reflex, bladder tone is greatly reduced. In performing the cystometrogram, a large volume of fluid could, therefore, be introduced into the bladder without the pressure rising significantly. When pressure did eventually start to rise, it is important to note that the patient was capable of voluntarily lowering this initial rise in pressure on request. This indicated that the patient must still have had an intact motor innervation to his bladder. Although its activity was minimal from the lack of the spinal sensory inflow to stimulate reflex motor outflow, sufficient spontaneous motor outflow persisted to summate with the intrinsic response of the stretched smooth muscle to produce some contraction with large distentions. The patient was capable of inhibiting this spontaneous motor outflow by descending inhibitory pathways from higher brain centers. With still further bladder distention, the bladder wall was stretched to the point where the connective tissue of the wall was bearing the tension, and neural inhibition could not alter this elastic tension. Under these circumstances of marked bladder distention, however, a normal individual would have experienced extreme discomfort as well as an urgent desire to micturate. The lack of discomfort in this patient at that point of the test corroborated his loss of sensory innervation from the bladder.

Micturition in the normal individual is accomplished by voluntary relaxation of the sphincters and shutting off the inhibitory suppression of the spinal reflex. The spinal reflex then stimulates contraction of the detrusor muscle of the bladder until bladder volume has been reduced to the point where the wall tension that is responsible for the spinal reflex stimulation becomes minimal. Even though this patient still seemed to have an intact motor innervation to his bladder, lack of the reflex drive prevented this motor innervation from effectively emptying the bladder. This patient was, therefore, obliged to rely on sphincter control and bladder compression by his abdominal muscles to evacuate the bladder.

This is a relatively ineffective method for emptying the bladder, and the patient still lacked any fullness signal to guide him in terminating his effort. A large residual volume of urine was, therefore, retained in the bladder. This urinary retention in a hypotonic bladder results in progressive chronic distention of the bladder. The cystitis that brought this patient to the clinic is a common complication in patients who retain a large volume of stagnant urine in their bladders.

Fluid, Electrolyte, and Acid-Base Regulation

VII

29

A 62-year-old woman was brought to the hospital complaining of 12 hours of vomiting and severe abdominal pain. Examination revealed an obese woman in acute distress with a blood pressure of 90/68 and a heart rate of 116 beats/min. Her abdominal wall was tense and very tender; and no bowel sounds were heard on auscultation. A laparotomy was performed and a twisted loop of bowel was found that was becoming gangrenous; 35 cm of the ileum was resected and intestinal continuity was reestablished.

The patient had a stormy postoperative course that required extensive medication. On the evening of the third postoperative day she had an attack of retching, during which her abdominal wound separated. She was returned to surgery and the wound was closed except for the insertion of a drainage tube. The patient was then started on an intravenous drip of 1 liter of 10% glucose in distilled water. At that time her blood pressure was 134/85.

At midnight the nurse called the house physician because the IV had clotted. The physician spent 30 minutes attempting to reestablish the intravenous infusion, but all available superficial veins appeared to have thrombosed as a result of previous venipunctures. He reasoned that the intestinal cavity contained a large surface area for fluid absorption and, therefore, instilled 1 liter of 10% glucose in water through the abdominal drainage tube.

Two hours later the nurse found the patient unresponsive, with a blood pressure of 55/?. A surgeon was called, who did an emergency cutdown to establish a route for intravenous therapy, and a transfusion of plasma was administered. The patient's blood pressure was promptly restored and she went on to full recovery.

Questions

Outline the physiological basis of the signs that dictated the initial surgical exploration.

What is the reason for administering 10% glucose in water to a postoperative patient?

Explain the mechanisms that would have precipitated shock following the instillation of the glucose solution into the abdomen.

Analysis of Case 29

The absence of bowel sounds indicated reflex inhibition of intestinal motility due to an acute focus of tissue damage in the intestinal tract.

The rigid abdomen is another reflex manifestation, indicating that the inflammatory process had extended to produce intense irritation of the parietal peritoneum. In addition, the hypotensive state of the patient was somewhat ominous. Although any severe visceral pain may induce hypotension, major disease in the distribution of the splanchnic circulation is especially prone to produce shock. Thus these signs clearly defined a surgical emergency.

From the start of premedication until full recovery from anesthesia, surgical patients have no fluid intake. Yet throughout this period, fluid loss continues through the skin, the respiratory tract, and the kidneys. Also, surgical exposure of visceral structures adds substantially to the fluid loss as a result of evaporation, and major surgery involves an appreciable loss of blood and tissue fluids. Similarly, surgery interrupts food intake, especially in cases involving the gastrointestinal tract. Thus 10% glucose in water postoperatively is a valuable therapeutic agent — the glucose to supply calories and the water to compensate for the loss of fluids, which, for the most part, are hypotonic losses.

The house physician was correct in recognizing that the peritoneal surfaces represent a large area for diffusion, but he forgot that these surfaces are a two-way street. When administered as a slow intravenous drip, solutions of glucose in water rapidly become hypotonic as the cells abstract the glucose. When instilled as a bulk solution into the abdominal cavity, however, a 10% glucose solution has twice the isotonic concentration and will initially draw fluid from the tissues into the abdominal cavity. Even more important, this liter of salt-free solution will cause the rapid diffusion of salt into the cavity, which will draw additional water along with the salt. The consequence of the abdominal instillation of this fluid will, therefore, be a sudden outpouring of extracellular fluid volume into the peritoneal cavity, which will be derived most directly from the plasma volume that brings this fluid to the abdominal organs. The consequent acute depletion of the plasma volume will plunge the patient into a state of hypovolemic shock. (This procedure of administering 10% glucose solution into the abdominal cavity has actually been used in laboratory animals as an experimental model of shock.) The administration of plasma, which contains an oncotic agent to ensure that the administered fluid would be retained in the vascular compartment, corrected this hypovolemia.

30

A 7-year-old child was brought to the hospital because of a convulsive seizure. By the time the patient arrived, he was comatose. The parents explained that the child had complained of severe constipation for the preceding 36 hours and had been unable to eat. They had attempted to treat the constipation with repeated tap-water enemas.

The patient had a blood pressure of 128/88 and a pulse rate of 85 beats/min. Except for noting evidence of compacted stools in the colon, the physical examination was negative. A catheterized urine sample was free of glucose and protein and had an osmolality of 685 mOsm. The laboratory reported the values for blood chemistry as follows:

Na	104 mM	Cl	58 mM
K	5.2 mM	HCO_3	16 mM
H^+	48 nM	Plasma osmolality	234 mOsm
Glucose	55 mg/100 ml		

The child was treated with intravenous fluids, consisting of 1.8% NaCl in 10% glucose solution, and demonstrated remarkable improvement.

Questions
What is the essential nature of this child's problem?

Explain its etiology.

What control mechanism is failing to function appropriately?

Analysis of Case 30
This is a case of water intoxication that is directly attributable to the use of tap-water enemas, free of any significant crystalloid. This hypotonic solution was absorbed by the intestinal mucosa and drastically diluted this child's body fluids.

The acid-base disturbance can be attributed to a mild ketosis associated with a lack of food intake for the previous 36 hours, with an increased anion gap of 35 mM, and there was hypoglycemia associated with this inanition.

The physiologically conspicuous finding was that the child was putting out a significantly hypertonic urine that would aggravate the problem. This result contrasts with the expected response to both the osmotic dilution of body fluids and the increased fluid volume in the body, either of which should have acted to shut off the release of ADH and to promote the excretion of a highly dilute urine to eliminate the excess load of water in the body.

Since cells are freely permeable to water, the extracellular hypo-osmolality would be accompanied by an equivalent intracellular hypo-osmolality, with a consequent swelling of cells. This cellular swelling leads to major disturbances in cerebral function, because neural function is very sensitive to transmembrane ionic balances and because the cerebrum is encased in the rigid cranium, which would cause cerebral pressure to rise as cellular swelling occurred. Nonspecific disturbances in cerebral function are typically characterized by a combination of excitatory and inhibitory manifestations, as indicated in this child by the earlier convulsion that was followed by a state of coma. A common excitatory phenomenon observed with brain-cell overhydration is activation of the supraopticohypophyseal system to release antidiuretic hormone. It is conceivable that there are distortions of the osmoreceptor membranes in a swollen brain that are analogous to membrane distortions produced in the receptor by hypertonic fluids; there are other known examples of this type of paradoxical response of receptor membranes. This ADH secretion was obviously inappropriate and served to intensify and protract the osmotic problem.

Prompt treatment of such cases with hypertonic solutions will osmotically counteract this swelling of cerebral cells and restore the function of the osmoregulatory system.

31

A 47-year-old male parking lot attendant was brought to the hospital in a stuporous condition, exhibiting muscular twitching that was initially interpreted as impending delirium tremens. His pulse was 88 beats/min, blood pressure was 120/70, and respiratory rate was 20 per minute. Nothing of significance was detected on physical examination except for hyperactive reflexes. His hemoglobin was 17.1 gm/100 ml with a hematocrit of 52%. The values obtained for his admission blood chemistries are as follows:

Na	143 mM	Cl	46 mM
K	2.5 mM	HCO_3	62 mM
H^+	25 nM	Arterial PCO_2	64 mm Hg
Arterial PO_2	58 mm Hg		

The patient was treated with 1 liter of 2.5% glucose in 0.5% saline to which 40 mM of potassium was added.

The next morning the patient remained severely depressed with sluggish and obtunded responses. He passed a dark stool, which laboratory tests confirmed as containing blood. The urine at this time had an H^+ concentration of 85 nM and a K^+ concentration of 65 mM. A repetition of his blood chemistries gave the following data:

Na	138 mM	Cl	64 mM
K	1.7 mM	HCO_3	56 mM
H^+	27 nM		

Over the next 24 hours the patient was given a total of 4 liters of glucose in 1% saline to which 120 mM of potassium was added.

The next morning the patient was remarkably improved; his plasma potassium was up to 3.1 mM and his plasma bicarbonate was down to 32 mM. He was excreting large volumes of urine with an H^+ concentration of 12 nM. His hemoglobin was 13.7 gm/100 ml with a hematocrit of 35%; BUN was 26 mg/100 ml. A repeat of his arterial blood gases showed a PO_2 of 86 and a PCO_2 of 48 mm Hg.

For the first time it was possible to obtain an accurate history. The patient claimed that he had never previously touched alcohol. Following the recent death of his mother, however, he had become so despondent that he had confined himself to the house with a large supply of liquor, which he consumed over a three-day period. He remembered little of that period except that the liquor made him very sick and he vomited a great deal.

An x-ray examination of his gastrointestinal tract was reported as normal; no further evidence of blood in his stools was found. He was discharged two days later.

Questions

What was the major disturbance in this patient that was demonstrated by the blood chemistry at the time of admission?

What two additional disturbances in acid-base chemistry does this patient exhibit?

Why did he fail to respond to treatment for the first 12 hours?

Analysis of Case 31

The dominant feature of the blood chemistry observed at the time of this patient's admission to the hospital was a severe metabolic alkalosis, as evidenced by the low H^+, the high HCO_3, and the elevated PCO_2. The muscle twitching observed was probably a result of this alkalosis. The acid-base picture, however, shows two additional complications. Cations totaling 145.5 mM were paired with measured anions of 108, giving an abnormally high anion gap of 37.5. This demonstrates the presence of a superimposed metabolic acidosis, presumably reflecting some keto-acidosis because of the period of inanition that the patient had experienced. In addition, the arterial CO_2 is somewhat more elevated than would be anticipated from the normal respiratory compensation to a metabolic alkalosis. This elevation, together with the arterial hypoxemia, indicates some impairment of pulmonary gas exchange, which would add a respiratory acidosis to the picture. The basis for the respiratory problem was never clearly established, but it was assumed to relate to some aspiration of vomitus at some point during his alcoholic indulgence.

Another aspect of the patient's initial condition was a severe dehydration, indicated by the high hematocrit and elevated hemoglobin — evidence that assumed greater importance the following morning when it became apparent that this hemoconcentration existed in spite of bleeding into the gastrointestinal tract. It should be noted that the patient's blood pressure on admission was somewhat lower than it was observed to be later. Reflex vasoconstriction can significantly restore blood pressure toward normal levels in spite of very substantial losses of fluid volume.

This state of dehydration apparently constituted the major drive on the kidney to promote sodium reabsorption, acting through the renin—angiotensin system to stimulate aldosterone release. At the same time, renal blood flow was undoubtedly favoring the medullary nephrons, which are the most powerful salt retainers. This avid salt retention under aldosterone stimulation would have favored renal loss of potassium, as was demonstrated by the urinary findings; and this renal loss of potassium aggravated the state of potassium depletion that had been produced earlier by the vomiting episodes.

The minimal correction of the alkalosis during his initial period in the hospital related to several factors that favored bicarbonate retention. The elevated CO_2 tension would have had a mass action effect on the renal tubules to accelerate the formation of carbonic acid, which increases the reabsorption of bicarbonate. At the same time, the avid reabsorption of Na cations demanded that equivalent anions must be reabsorbed, and the hypochloremia provided a limited amount of chloride to accompany the sodium. This imbalance created additional pressure for bicarbonate reabsorption. Finally, an alternative method for electrochemical balance in sodium reabsorption would involve cation exchange across the tubule: The depleted potassium supply would encourage exchange of H^+ for Na^+. This exchange was occurring at an appreciable rate, as demonstrated by the H^+ being excreted in the urine; for each H^+ ion excreted, an HCO_3^- ion was returned to the plasma. Renal function was, therefore, acting to worsen rather than to correct the alkalotic state.

To reverse this process, it was essential to provide ample isotonic fluids to restore the extracellular fluid volume and to suppress the action of the control mechanisms acting to conserve sodium, to restore the chloride level to encourage the excretion of bicarbonate anion, and to replenish the potassium levels to foster an exchange of Na^+ for K^+ instead of H^+. With these corrections achieved, the kidney began putting out copious volumes of alkaline urine to correct the electrolyte disturbance.

No further evidence of gastrointestinal bleeding was observed in this patient. The bleeding was probably due to an alcoholic gastritis that resolved when the irritant was removed.

An 8-year-old girl was brought to the hospital with the history of four days of profuse diarrhea. She was listless and responded rather incoherently to questions. Her skin turgor was poor for a child of her age and her eye balls were soft and sunken. Her pulse was 114 beats/min with a blood pressure of 98/66. Her respirations were deep and at a rate of 26 per minute. Her hematocrit was 58%. Her lungs were clear and her abdomen was soft and without evidence of significant local tenderness. A blood sample was taken; part of it was used for a determination of hydrogen-ion concentration, and the remainder was sent to the laboratory for a complete analysis. When the physician discovered a plasma hydrogen-ion concentration of 74 nM, he ordered an intravenous infusion of 700 ml of 5% glucose, containing 10 mM of $KHCO_3$ and 110 mM of $NaHCO_3$.

One hour later the nurse discovered that the patient had stopped breathing. No pulse was detectable and auscultation of the chest revealed weak heart sounds at a rapid rate. Attempts to resuscitate the child were unsuccessful.

A call to the laboratory revealed that on admission the child had a plasma HCO_3 of 6 mM, a P_{CO_2} of 18 mm Hg, and K^+ of 5.8 mM.

Questions

What was the nature of this child's problem on admission?

Why was K^+ ordered to treat this condition?

What is the danger in the therapy that was employed?

Analysis of Case 32

This child was admitted to the hospital with a severe metabolic acidosis and dehydration due to the extensive loss of bicarbonate and fluid in the diarrhea. Therapy in such a case must be primarily directed toward correcting the dehydration, which will provide some amelioration of the acidosis by dilution as well as provide the kidney with the excess fluid and electrolyte it needs to readjust the acid-base balance (cf. Case 31).

These patients also confront the problem of a potassium deficiency, a factor that must be recognized if specific treatment of the acidosis is employed. Intestinal secretions contain significantly more potassium than the extracellular fluid. In normal body economy, these secretory products are reabsorbed at lower levels of the tract and recycled without a significant net loss from the body. Diarrhea breaks this cycle and leads to significant potassium losses. Superimposed on this loss is the role of cellular and skeletal elements in buffering hydrogen ion. As loss of bicarbonate leads to increased hydrogen ion in the extracellular compartment, ionic equilibria of both the intracellular phase and the exchangeable cation sites on the bone mineral are shifted, with hydrogen ion moving out of the extracellular compartment and being replaced with potassium ion. The kidneys respond to this influx of potassium ion into the plasma and increase potassium excretion, adding urinary losses of potassium to the losses in the diarrheic fluid.

Because fluid and potassium are being lost simultaneously, plasma concentrations of potassium are of little help in assessing this problem. In this patient, for example, the observed potassium level could be considered quite ample if the mistake of relating plasma concentration to total body potassium is made. Rapid rehydration of such a patient with potassium-free solutions will lower this plasma potassium concentration, and correction of the acidosis will reverse the buffering process with uptake of potassium by the cells and bone minerals. This therapy can produce a precipitous drop in plasma potassium concentration. Hence the rather conservative dose of potassium administered to this patient would serve to attenuate this drop in extracellular potassium concentration.

Rapid correction of a metabolic acidosis encounters the even greater danger of producing acute cerebral acidosis. The cerebrospinal fluid, like the systemic body fluids, relies upon the bicarbonate system as its major buffer for hydrogen ions. Although CO_2 diffuses readily across it, the blood-brain barrier is impermeable to both H^+ and HCO_3^- Cerebrospinal fluid buffering is controlled independently by bicarbonate "pumps," which actively transport HCO_3^- either into or out of the CSF to control its hydrogen ion concentration. Because the cerebral bicarbonate reservoir is relatively large in comparison with the limited capacity of the bicarbonate pumps, however, it takes many hours for this system to respond to an abrupt change in acidity. For the body as a whole, the immediate physiological defense against a metabolic acidosis is stimulation of the carotid and aortic chemoreceptors by the increased H^+ in the plasma to increase the respiratory drive. The resulting hyperventilation lowers the plasma CO_2 and, consequently, the carbonic acid concentration in the blood to blunt the rise in plasma hydrogen ion. At the same time, however, this lowered CO_2 equilibrates across the blood-brain barrier and similarly lowers the H^+ in the CSF. Since in many instances the brain is not involved in the disturbance responsible for the systemic acidosis, the H^+ in the CSF often drops below normal during the early phase of a metabolic acidosis. This alkalotic swing of the CSF stimulates the HCO_3^- pumps to actively transport bicarbonate from the CSF to the plasma compartment. At the time this child was admitted to the hospital, this process had been going on for approximately four days, ample time for the CSF bicarbonate to have been lowered to match the low CO_2 and achieve normal acidity of the fluid bathing the brain.

The bicarbonate administered in the hospital would have been distributed throughout the extracellular volume and, to a limited extent, it would equilibrate with the intracellular phase, but it would not cross the blood-brain barrier. Assuming a bicarbonate space of 40% of body weight and approximating the weight of this patient as being on the order of 30 kg, the 120 mM of bicarbonate should have raised her plasma bicarbonate concentration by 10 mM. This 10 mM elevation might appear conservative for the degree of acidosis that she demonstrated. The bicarbonate was infused directly into the blood stream, however, so that the acidity of her arterial blood would have been corrected well before this whole body equilibration would have been achieved. Lowering the H^+ in her arterial blood would have eliminated the stimulation of the chemoreceptors, slowed respiration, and elevated her CO_2 tension. This elevated CO_2 would have diffused rapidly across the blood-brain barrier into the bicarbonate-depleted cerebrospinal fluid and significantly increased the acidity in the CSF. Rapid correction of a severe metabolic acidosis that has persisted for some time, therefore, confronts the hazard that, in correcting the systemic acidosis, an acute cerebral acidosis will be produced. This can prove fatal, as in the case of this child.

When this hazard was first appreciated, the notion arose that protection from catastrophe would be afforded by the use of such alkalinizing salts as sodium lactate in preference to sodium bicarbonate to delay the speed of the alkalinization. It has come to be recognized, however, that tissues such as the liver and the heart very rapidly convert lactate into bicarbonate so that no significant advantage is gained by the use of alternative agents. Rather, therapy must be managed with this danger in mind, resorting to bicarbonate administration only when the acidosis is deemed life-threatening, and watching the patient very carefully for signs of cerebral failure.

33

A 54-year-old lawyer in a stuporous condition was admitted to the hospital via the emergency room. His wife reported that he had become increasingly lethargic during the preceding 12 hours. She stated that two years earlier her husband had been told that a routine urine examination revealed protein in the urine, but her husband had rejected any follow-up because he had felt in good health.

The patient's blood pressure on admission was 194/122 with a pulse rate of 80 beats/min. His respiration was 24 per minute. A catheterized urine specimen was cloudy with a 4+ albumin and a hydrogen-ion concentration of 14 nM; the urine contained no glucose. Blood chemistries on admission were as follows:

Na	135 mM	Cl	128 mM
K	5.7 mM	HCO_3	8 mM
H^+	66 nM	PCO_2	22 mm Hg
BUN	62 mg/100 ml	Hemoglobin	15.2 gm/100 ml
Creatinine	8.4 mg/100 ml	Hematocrit	46%

The patient was treated with parenteral fluids that contained bicarbonate and was given a drug to lower his blood pressure. The next day he was responsive and reported that he felt much better. A 24-hour urine collection demonstrated a creatinine excretion of 1.6 gm; his creatinine level on that day was 8.2 mg/100 ml.

Questions

What condition brought this patient to the hospital?

Explain his acid-base disturbance.

Why would such a patient not show glucose in the urine?

Analysis of Case 33

This patient suffered from chronic renal disease that he was able to ignore until he encountered this severe bout of renal acidosis. A renal basis for the acidosis is demonstrated by the hyperchloremia and the absence of an abnormally high anion gap, excluding the metabolic production of acid as a cause for the acidosis. It is particularly significant that this patient's urine was alkaline in spite of his severely acidotic state, which indicates spillage of bicarbonate by functionally impaired renal tubules. This condition contains a vicious-cycle component, in that the excretion of bicarbonate leads to chloride retention to maintain electrochemical balance with the reabsorbed sodium; and the resulting hyperchloremia of the filtrate provides a large supply of anions for reabsorption, which compromises bicarbonate reabsorption even more. At the same time, the hyperventilation induced by the metabolic acidosis acts to lower the PCO_2 in the renal tubular cells, which depresses the synthesis of carbonic acid and interferes with bicarbonate reabsorption still further. Retention of other acid anions (PO_4, SO_4) by the failing kidney will contribute to the acidosis, but these acids are less important quantitatively, as indicated in this case by the small anion gap.

The tubules responsible for the loss of bicarbonate in this patient, however, must have represented a small fraction of his total renal tubules. His creatinine clearance was 19 liters/day, or 13.6 ml/min, which is roughly one-tenth of the normal value. Therefore some 90% of this man's renal tubules were completely nonfunctional since they were passing no filtrate. With such a severe reduction in filtration, renal glucose reabsorption was undoubtedly very severely reduced. The few tubules that were still functional were capable of handling the small load of glucose presented to them; glucose that was not filtered could not appear in the urine. Tubular reabsorptive capacity for glucose can only be assessed by comparing the filtered load with the glucose excreted. Thus the absence of glucose in the urine provides little assurance that renal function is normal.

The urea that builds up in the blood of patients with failing kidneys has no inherent toxicity, but it does diffuse readily across biological membranes including the intestinal mucosa. The intestinal flora normally include organisms with very high urease activity, capable of converting urea to ammonia. If this ammonia diffuses back into the blood faster than the liver can recycle the ammonia back to urea, blood ammonia levels will rise. Elevated ammonia levels disrupt brain metabolism and contribute to the coma observed in renal or hepatic failure. (There are additional factors contributing to the coma produced by a renal acidosis, however, as discussed in Case 37.)

34

A 46-year-old salesman, who was a heavy cigarette smoker with a long history of bronchopulmonary disease, was admitted to the hospital with cyanosis, dyspnea, and fever. His blood pressure was 138/78, his pulse was 95 beats/min, respiration was 34 per minute, and rectal temperature was $43.0°C$. The patient periodically coughed up tenacious purulent sputum. Examination of the chest revealed rales and wheezes throughout all lung fields. The laboratory data obtained on admission were as follows:

Hemoglobin	17.1	Hematocrit	53%
Na	143 mM	Cl	78 mM
K	4.8 mM	HCO_3	38 mM
H^+	46 nM	PCO_2	73 mm Hg
PO_2	47 mm Hg		

The patient was treated with antibiotics and expectorants to loosen his sputum. He was also encouraged to cough to raise as much sputum as possible. The next morning the patient felt much better, and his temperature was down to $37.8°C$. His cyanosis had disappeared and he was breathing easily. Breath sounds demonstrated only an occasional crackle or wheeze. Repeat blood chemistries that morning were as follows:

Na	141 mM	PCO_2	44 mm Hg
K	4.6 mM	Cl	84 mM
H^+	34 nM	HCO_3	31 mM

Questions
Characterize the acid-base disturbance in this patient on the night of admission.

By contrast, what was his acid-base status the next morning?

Why was oxygen therapy not used?

Explain how blood chloride could have risen during a period when there was no ingestion of chloride.

Analysis of Case 34

On admission this patient had a respiratory alkalosis due to major airway obstruction and alveolar hypoventilation. His acute state was precipitated by a bout of bronchopneumonia. This episode, however, must have been superimposed on a severe degree of chronic hypoventilation in view of the substantial elevation of his bicarbonate. The latter reflected increased renal reabsorption of bicarbonate in the face of a sustained hypercapnia. This elevated CO_2 tension in the renal tubule cells increased their synthesis of carbonic acid and thereby provided the substrate for greater bicarbonate reabsorption; this served a compensatory function in curbing the rise in the hydrogen-ion concentration due to his respiratory insufficiency.

The patient showed a dramatic response to his antibiotic therapy and steps to clear his airways. The next morning he, therefore, exhibited a condition that would be classified technically as a metabolic alkalosis. The relief of his airway obstruction improved gas exchange, eliminating the cyanosis and causing a significant drop in his P_{CO_2}. The large reservoir of bicarbonate that had been accumulated in his body fluids, however, could only be excreted slowly in the urine. A factor that was "compensatory" initially had now become the prime disturbance. The resulting fall in plasma hydrogen ion removed a stimulatory drive on the chemoreceptors and reduced respiration. Thus the residual elevation of his CO_2 tension can be considered compensatory to his metabolic alkalosis.

As soon as the P_{CO_2} fell, the renal tubules would slow down their synthesis of carbonic acid entering into the exchange reaction with the bicarbonate in the tubular lumen, with the result that the kidney would start to spill bicarbonate. This bicarbonate in the urine would carry an equivalent amount of water with it osmotically. The extracellular fluids were, therefore, being depleted of both bicarbonate and water, while chloride would be conserved as the major anion accompanying reabsorption of the filtered sodium. Thus the rise in plasma chloride reflects, in part, a loss of extracellular water without an equivalent loss of chloride. In addition, the reduction in the plasma bicarbonate concentration would shift the equilibrium across the red cell membrane and cause the movement of a significant amount of chloride from the red cells to the plasma. If this patient's pulmonary function could be restored to the point where no further CO_2 retention occurred, this adjustment would continue until the chloride level had been raised and the bicarbonate level had been lowered to normal levels. With the degree of chronic pulmonary insufficiency the patient demonstrated on admission to the hospital, however, it is probable that he will continue to exhibit significant elevations in bicarbonate to maintain reasonable compensation, keeping his hydrogen-ion concentration fairly close to normal, in spite of his chronic state of respiratory acidosis.

35

A 5-year-old boy was brought to the emergency room in a somewhat stuporous condition and exhibiting evidence of delirium. His parents reported that the boy had been in good health until 2 hours before coming to the hospital. At that time he had vomited, after which he had shown progressive signs of deterioration of his mental state.

Nothing of significance was noted on examination of the child except that his respiration was rapid and deep. Determination of the plasma bicarbonate in the emergency room gave a value of 16 mM.

The child was admitted to the hospital and complete blood chemistries were ordered. The significant findings were: H^+, 18 nM; HCO_3^- 13 mM; and PCO_2, 10 mm Hg. The physician left orders to have the child watched carefully for any change in his condition but with no specific therapy.

Over the next 24 hours the child's mental state gradually cleared. Blood chemistries were repeated the following day and the following data were obtained:

Na	142 mM	Cl	84 mM
K	5.8 mM	HCO_3	19 mM
H^+	48 nM	PCO_2	35 mm Hg

Questions

What medical emergency is most commonly associated with coma and a low plasma bicarbonate?

What would have been the consequences of giving this child specific therapy to combat this more commonly observed acid-base disturbance?

What type of disturbance was demonstrated by the more complete blood chemistry obtained on the first day after admission to the hospital?

Of what diagnostic significance was the increased respiration noted in the emergency room?

What type of disturbance was present on the second day?

Analysis of Case 35

This case dates back to a period when emergency rooms were not equipped to measure hydrogen-ion concentrations; the only quick diagnostic aid available to the physician for identifying an acid-base disturbance was a technique that measured plasma bicarbonate levels. A low bicarbonate level is a sign of either a metabolic acidosis or a severe respiratory alkalosis. If a physician confronting a comatose child with a low bicarbonate level assumed that he was dealing with a metabolic acidosis, and if the panic of the patient's distraught parents encouraged him to treat this assumed condition vigorously, the treatment could be fatal. Many children with the same problem that existed in this boy have been given bicarbonate infusions to correct a supposed metabolic acidosis in this fashion, and the tragic number of fatalities that have resulted served as a major stimulus to include hydrogen-ion determinations as part of the armamentarium of the emergency room.

A valid diagnosis of the nature of the acid-base disturbance in this child was not possible without the report of the complete chemistries obtained after admission to the hospital. These values demonstrated a respiratory alkalosis, inasmuch as all three variables of the bicarbonate buffer equation were low. The augmented respiration of the child noted in the emergency room would have been of no help in making this diagnosis because a metabolic acidosis would also have produced increased respiration by chemoreceptor stimulation. The decision as to whether increased respiration is the primary disturbance or a secondary compensation cannot be made by observing the patient; it demands the chemical data from the laboratory. In this instance the data showed that the primary disturbance was a blowing off of CO_2 that lowered the concentration of hydrogen ion in the blood; the reduced CO_2 tension depressed carbonic acid synthesis in the kidneys secondarily and, therefore, the filtered bicarbonate was not being reabsorbed by this boy's kidneys. This condition had persisted long enough for this compensatory fall in bicarbonate to be evident in the plasma level, indicating that the boy had been hyperventilating for several hours.

The following day the pendulum had swung in the other direction with the hydrogen-ion concentration rising above normal, indicating that the child's condition had been transformed into a metabolic acidosis. This diagnosis was supported by the fact that measured cations totaled 147.8 mM while measured anions equaled only 103, leaving a large gap of unmeasured anions of 44.8 mM. Thus some new anion was being added to the system and displacing bicarbonate ion.

This child's problem was acetylsalicylic acid poisoning, attributable to the marketing of candy-flavored aspirin to make the drug more palatable for children. The maneuver was tragically successful in that the drug became so attractive that children who found a bottle of this candied aspirin consumed the contents of the bottle and acquired serious aspirin poisoning. Aspirin in high doses has a number of toxic actions on the nervous system, accounting for the signs and symptoms that brought this child to the hospital. The action of immediate importance to us is a direct stimulatory action on the respiratory centers, causing CO_2 to be blown off and thereby producing the respiratory alkalosis observed. A sustained respiratory alkalosis, particularly as observed in cases of aspirin poisoning, alters enzymatic activities and stimulates the production of a large amount of lactic acid. This accumulation of lactic acid accounts for the large anion gap and the metabolic acidosis observed in this child on the second day.

Consideration of currently accepted treatment for aspirin poisoning is beyond the scope of our concern. Suffice it to say that the physician in this case suspected the diagnosis and was apparently satisfied, from the parent's description of the vomiting episode, that further evacuation of the gastric contents of the child was not warranted. The physiologically important lesson to be learned from this problem is the danger of attempting to deal with acid-base disturbances without adequate data for understanding the nature of the condition that is being treated.

36

A 32-year-old male visited his physician complaining of frequent bouts of palpitation and excessive perspiration. Examination revealed a rather tense individual with a wet handshake, a pulse of 100 beats/min, and a blood pressure of 130/95. In the absence of other significant findings, the physician made arrangements for the patient to obtain a metabolism test. When his basal metabolic rate was reported back as 25% above normal, the physician advised hospitalization for a diagnostic workup.

On the afternoon of admission to the hospital, the patient appeared to be, generally, in good health. On a routine admission examination, his respiration was 16 per minute, his pulse was 95 beats/min, blood pressure was 146/98, hemoglobin was 15.2 gm, and urine showed a trace of sugar (1+) but no albumin. The patient was assigned a bed and scheduled for iodine-uptake studies the next day. At 2:00 A.M. a house physician was called by the nurse to examine the patient. The patient had vomited a large quantity of undigested food. He was perspiring profusely and the bedding was soaked with sweat. His pulse rate was 132 beats/min, and his blood pressure was 215/138. Examination of the abdomen revealed an absence of any bowel sounds; no evidence of tenderness was found. The bedding was changed, and the patient was given appropriate sedation.

The next morning the patient had a pulse of 82 beats/min and a blood pressure of 124/82, and he reported that he felt quite comfortable. Analysis of his urine was reported back as having 4+ glucosuria. On this basis, iodine-uptake studies were canceled and a blood glucose determination was ordered. His blood glucose was reported as 88 mg/100 ml. A consultation was then ordered and the patient was given a very thorough physical examination. No findings of significance were encountered, but it was noted at the conclusion of this examination that the patient had a blood pressure of 160/112. It was decided to perform some special tests on the patient's urine.

On the basis of these urinary findings, a definitive treatment was carried out.

Questions
What did the patient's physician originally suspect, and on what evidence was this suspicion based? Explain the detailed physiological basis for relating the observed signs and symptoms to this disease.

Explain the physiological mechanisms responsible for the signs and symptoms associated with the episode the patient experienced during his first night in the hospital.

What common denominator can you identify in this picture?

What could explain the glucosuria?

What compounds should be looked for in the urine?

Analysis of Case 36

This patient's case resembled one of hyperthyroidism when first seen. This resemblance is not coincidental, but reflects the fact that his true problem and many of the manifestations of hyperthyroidism are acting through the same mechanism. An important action of thyroid hormone is to augment the response of adrenergic effectors to norepinephrine, which is released by sympathetic nerve stimulation. This explains the tachycardia and, as will be discussed, is related to his increased perspiration. The moderate elevation of the basal metabolic rate accords with hyperthyroidism but could also be a manifestation of the calorigenic action of excess catecholamines. The only feature in the initial examination that argues against hyperthyroidism is the elevated diastolic pressure. The generalized hypermetabolism produced by excess thyroid hormone causes autoregulatory vasodilation to increase blood flow to the tissues, producing a lowered diastolic pressure, increased cardiac output, and a wide pulse pressure.

The episode during the patient's first night in the hospital was clearly that of an intense adrenergic activation. Vomiting undigested food a number of hours after eating is a clear sign of inhibition of the gastrointestinal tract, as was also evidenced by the lack of bowel sounds. The tachycardia and hypertension indicate an adrenergic effect. The glucose found in the patient's urine the next morning, unassociated with any sustained hyperglycemia, undoubtedly reflects a transient hyperglycemia during the attack at night.

The profuse perspiration often seen in this disease is the subject of considerable confusion because textbooks state rather too glibly that the sweat glands are cholinergic receptors. While it is true that the innervation of the eccrine sweat glands is cholinergic, these glands can also be stimulated by circulating norepinephrine. In addition, the conspicuous sweating of the hands includes an important apocrine component, which is a true adrenergic response. Finally, the calorigenic effect of the sudden release of catecholamines into the circulation could activate heat-loss mechanisms via cholinergic stimulation of the eccrine glands. Circulating norepinephrine acts synergistically with this cholinergic stimulation to augment sweat secretion.

A review of his picture, reinforced by the mild hypertensive response evoked by the physical examination, prompted a change in the diagnostic effort toward assessing the possibility that this patient was producing excess quantities of adrenal medullary hormones. While it is becoming possible to measure norepinephrine and epinephrine directly in the blood, at the time that this patient was being examined, the only reliable index of increased catecholamine production was to search for their metabolites in the urine: metanephrine and vanillylmandelic acid. Large amounts of these products were found in the urine of this patient, and he was explored surgically. It was his good fortune that the surgeon found a large solitary pheochromocytoma, which was removed, and the patient had an uneventful recovery.

37

A mother brought her 13-year-old daughter to the physician. The mother was obviously distraught over what she described as "emotional disturbances" of her daughter, characterized by bed wetting, pulling and scratching in the genital area, and generalized "laziness." Questioning the daughter revealed recently appearing habits of eating between meals and frequent fluid intake to satisfy thirst. Menstrual periods had begun six months earlier.

The child appeared as rather thin and frail, but otherwise a normal adolescent. The only significant finding on physical examination was inflammation of the vulva and deposits in the folds of the labia that were suggestive of a fungus infection. Urinalysis demonstrated glucosuria, and her blood sugar was 160 mg/100 ml. The diagnosis of diabetes mellitus was made and the patient was instructed in the use of insulin; after several months she was well stabilized on 70 units of insulin per day.

At the age of 16 the patient was admitted to the emergency room of the hospital in coma. The history obtained from the mother was that the girl had attended a school dance the previous evening and arrived home very late. Her mother had felt it important to let her sleep the next morning and had not attempted to awaken her until noon, at which time she was unable to arouse her daughter adequately.

Examination revealed a comatose patient with slow but very deep respirations. Her urine was positive for glucose and acetone bodies, and a plasma sample was noted to be milky in appearance. Data reported by the laboratory were as follows:

Na	152 mM	Cl	88 mM
K	7.1 mM	HCO_3	10.8 mM
H^+	54 nM	BUN	35 mg/100 ml
Glucose	312 mg/100 ml		

The patient was treated with insulin and intravenous saline. Initially, the patient showed some improvement, but 2 hours later she became stuporous again with a blood pressure of 80/50. At this time blood glucose was found to be 100 mg/100 ml, there were no ketone bodies, plasma H^+ had fallen to 42 nM, and K^+ had fallen to 2.2 mEq.

Questions

Which of the original presenting symptoms that the mother ascribed to "emotional disturbances" were related to the child's diabetes?

Explain the metabolic picture seen at the time of her hospital admission.

Explain the initial and the secondary electrolyte disturbance observed in this episode.

The patient exhibited severe depression of CNS function twice during the hospital admission. What physiological basis for these disturbances can you suggest?

Analysis of Case 37

The original history of this child demonstrates the spectrum of problems that are characteristic of the diabetic state. Loss of weight in the face of increased food intake indicates a lack of cellular utilization of ingested calories, a factor also contributing to the easy fatigability and lassitude that the patient's mother interpreted as "laziness." The unmetabolized glucose was spilling into the urine because filtered glucose exceeded the renal tubular reabsorptive capacity for glucose and acted as an osmotic diuretic as it traversed the collecting-duct system. The bed wetting was a manifestation of this diuresis, which obviously had to be coupled with increased fluid intake to maintain water balance. The appearance of the disease in this patient at the time of puberty was probably coincidental. On the other hand, animal models suggest that this type of "juvenile," or early-onset, diabetes could relate to a fundamental deficiency in the insulin-producing alpha cells, which are incapable of responding to some of the metabolic surges of the growing child. A deficient response of the alpha cells to the metabolic demands of adolescence could intensify and prolong the stimulation of a glucose load on these cells and lead to exhaustion atrophy of the alpha-cell tissue.

The infection of this patient's genitalia could relate to her diabetes in two ways. An obvious factor is the copious supply of nutrient being offered for fungal growth by the glucose in the urine. In addition, the lack of cellular uptake of glucose leads to increased secretion of glucocorticoid, which has a suppressive action on the inflammatory responses that help to combat such infections.

The patient was admitted to the hospital showing the characteristic picture of a diabetic acidosis with hyperglycemia, low bicarbonate, and a compensatory hyperventilation. Though blood levels of keto-acids were not reported, their presence is evident in the large anion gap, the measured cations exceeding the measured anions by 60 mM. The milky plasma would indicate hyperlipemia, which is caused by a failure to clear the chylomicrons that are formed from ingested fat when there is a failure of activation of the lipoprotein lipase in the absence of insulin. The elevated blood urea nitrogen is a by-product of the greatly accelerated gluconeogenesis from amino acids, which is exaggerated by the renal failure as dehydration reduces blood pressure and glomerular filtration.

Questioning the patient after she had recovered from the coma episode revealed that in the excitement of anticipation of the school party, she had omitted her insulin injection. The intensity of the metabolic derangement that this deficiency produced was undoubtedly exaggerated by the sympathoadrenal stimulation associated with the excitement of the party.

The coma that is characteristic of severe diabetic acidosis appears to represent the summation of a number of factors. Brain function is dependent on (1) the ionic balance across the cell membranes, (2) the mechanisms for the transient changes in these balances associated with cell excitation, and (3) maintenance of the energy supply to sustain the first two factors. Acidosis directly disturbs the first two factors. Intracellular anions have dissociation constants for hydrogen ion that make them good buffers. As these anions combine with hydrogen ion, however, an equivalent amount of potassium cation will no longer be held within the cell and, therefore, will diffuse into the extracellular compartment. This shift of K^+ from the intracellular to the extracellular phase reduces the membrane polarization. Though transient reductions in membrane polarization increase cell excitability, sustained depolarizations act to raise the threshold and decrease cell excitability. In regard to the actual activation process associated with cell excitation, generation of the ionic currents to produce action potentials is very sensitive to the external concentration of calcium ion, being blocked by excessive calcium ion and developing spontaneously (tetany) with a lowering of the calcium ion level. The ionization of calcium is, in turn, dependent on the acidity, increasing as H^+ ion increases. This increased ionized calcium interferes with the opening of membrane channels for sodium diffusion that is essential for cell excitation. These two independent actions of acidity on K^+ and Ca^{++} account for a universal tendency for acidosis to depress neural function.

It is very doubtful, however, whether diabetic coma can be simply characterized as an acidotic coma. The diabetic becomes comatose at significantly less severe states of acidosis than is observed in other types of acidotic coma. Furthermore, it must be recognized that brain metabolism is not as dependent on insulin as are peripheral tissues. Diabetic acidosis is caused by the generation of keto-acids because of the overwhelming breakdown of fatty acid in the liver. Due to the strong chemoreceptor stimulation that results from the presence of these organic acids in the arterial blood, brain CO_2 tension is significantly lowered. The degree of cerebral acidosis in the diabetic is, therefore, much less than the acidosis observed in the blood.

A factor totally unrelated to the acidosis is the hyperosmolarity that develops in the uncontrolled diabetic. This is partially attributable to glucose itself, since each increment of 100 mg/100 ml of glucose increases osmolarity by 6 mOsm. In addition, the osmotic diuresis produced by the glucose preferentially eliminates water with retention of salt, frequently producing a hypernatremia, as was observed in this case. Hyperosmolarity, therefore, directly disturbs ionic concentrations. In addition to ionic disturbances, hyperosmolality in itself can disturb neural function to the point of producing coma, as is occasionally observed in the uncontrolled diabetic whose hyperglycemia exceeds 1000 mg/100 ml in the absence of significant acidosis or hypernatremia.

Finally, the metabolic energy supply to the brain, as judged by cerebral oxygen consumption, is severely depressed in diabetic coma, more so than in other forms of coma. While nerve cells are not dependent on insulin for the transport of glucose into the cell, there is evidence that high concentrations of acetoacetate have a specific inhibitory effect on energy production by the Krebs cycle. It is evident that diabetic coma does not have a simple explanation. Presumably, the interaction of these several factors summates to produce the comatose state in the diabetic.

The secondary depression of the central nervous system observed in this patient, following her initial therapy, was related to another action of potassium. Supplying the glucose-starved cells with insulin restores the balance of carbohydrate and fat metabolism and shuts off the production of keto-acids. Also, the availability of glucose to the cells restores concentrations of some of the phosphorylated anions within the cells. Both these factors result in a dramatic shift of K^+ from the extracellular to the intracellular compartment, as evidenced by the drop from 7.1 to 2.2 mM in the plasma potassiums recorded for this patient. This shift of K^+ produces hyperpolarization of cells, which also serves to reduce their excitability and depress neural function. Potassium therapy, therefore, is an important adjunct in the treatment of diabetic coma but, obviously, it must be employed judiciously. As was true in this case, prior to insulin therapy the potassium level may be running dangerously high; hence the therapeutic regimen must be adjusted to provide the potassium at the appropriate phase of the therapeutic response to insulin.

38

Note: This was a further episode in the history of the same patient described in Case 37.

A 19-year-old college student was brought to the hospital in a comatose state after she collapsed at a bus terminal. A traveling companion explained that they were returning home from a semester at college, where they had completed final examinations. They had just disembarked from a 2 hour trip on a bus, which they had been able to catch by omitting lunch. On alighting from the bus, the patient became confused and disoriented and staggered to the floor. The companion reported that she understood that her friend was a diabetic.

The condition of the patient while she awaited an ambulance at the bus terminal was described as delirious; her pupils were widely dilated and the eyeballs were twitching; her skin was very pale and she was perspiring profusely. She had a strong and rapid pulse. At intervals she exhibited mild convulsive episodes.

On admission to the hospital, blood and urine samples were immediately obtained, and the patient was given 10% glucose solution parenterally. Her blood sugar was found to be 48 mg/100 ml and her urine was negative for sugar and acetone. The patient responded promptly to the therapy.

Questions
What is the physiological explanation for the attack the patient experienced at the bus terminal?

What influence does such an attack have on her immediate physiological condition?

Is it wise to give a diabetic in a comatose state intravenous glucose without laboratory data to confirm a suspected diagnosis?

Analysis of Case 38
Blood sugar is regulated by a complex of hormonal controls, with glucocorticoids, glucagon, growth hormone, and catecholamines all acting to elevate blood sugar, and insulin acting to depress blood-sugar levels. Insulin deficiency deprives the patient of the means for counterbalancing these other control mechanisms. While, theoretically, one may regard the administration of insulin to the insulin-deficient form of diabetes as a "cure" of the disease, in reality, this therapy falls signifi-

cantly short of being a cure because the artificially injected hormone is not subject to the feedback regulation of the endogenous hormone secretion.

College examination periods are emotionally stressful experiences for students, complicated by disturbances in their eating and sleeping patterns and by exaggerated activity levels of both their somatic and autonomic nervous systems. This places a heavy burden on mechanisms for blood glucose regulation. During the examination period itself, catecholamines will tend to be at a high titer, mobilizing blood glucose by glycogenolysis in the liver, as well as stimulating the release of free fatty acids from adipose tissue. These energy substrates will be rapidly taken up by peripheral tissues, especially muscle, since the nervous tension caused by a period of stress is manifest by very substantial increases in muscle tone. This patient's rush to catch the bus after her examinations would represent a further bout of muscle exercise and autonomic hyperactivity. During the 2 hour bus ride, this situation was totally reversed. Relaxation from the stressful period, while riding on the bus, would result in a profound depression of her heightened autonomic activity and disappearance of the catecholamine stimulus for mobilizing blood sugar. Skeletal muscles would be in the process of restoring their energy reserves that were depleted by their previous hyperactivity. Lack of alimentary intake because of the omitted lunch would provide no support for sustaining blood-sugar levels during this period. In the normal individual, the tendency toward hypoglycemia in this situation is blunted by shutting off the secretion of insulin. In this patient, her injected insulin was continuing to be released without any feedback control. Her blood sugar, therefore, fell to low levels because of this uninhibited release of insulin.

While most tissues of the body can make substantial use of fatty acids as a substrate for their energy requirements, nerve tissue is uniquely dependent on glucose for its energy. In chronic hypoglycemic states, metabolic pathways in the brain can adapt to sustain energy supplies at remarkably low levels of glucose. In the diabetic, who would more likely exhibit metabolic adaptation to hyperglycemic states, normal brain function cannot be preserved if blood sugar falls below levels of approximately 50 mg/100 ml. The brain has a built-in safety mechanism, however, to protect against hypoglycemia. When glucose supply is restricted to a region of the brain stem, a massive discharge of the sympathoadrenal system is triggered off. This discharge provides the blood sugar mobilizing actions of the catecholamines to counteract the hypoglycemia. Most of the manifestations of the insulin shock exhibited by this patient should be recognized as due to intense activation of the sympathetic nervous system, augmented by some convulsive manifestations as depressed higher brain centers release their control of subcortical systems.

Since the blood-sugar elevations produced by the insulin shock responses are relatively transient, this condition urgently demands prompt action to correct the hypoglycemia. Because of the difficulties of obtaining a history from a disoriented or comatose patient, it is not always easy to be certain whether such an episode is a manifestation of hypoglycemia or of hyperglycemic acidosis. The hyperglycemia itself, however, is a physiologically innocuous feature of uncontrolled diabetes until exceedingly high levels of blood sugar are achieved. The relatively modest elevation of blood-sugar levels produced by parenteral glucose therapy, therefore, does not significantly influence the condition of the patient in diabetic coma, while it can be life-saving for the patient in hypoglycemic shock.

39

A 42-year-old woman was admitted to the hospital complaining of abdominal pain. She stated that she had felt poorly for the past year, having been bothered by constipation, loss of appetite, and lethargy.

She was an obese woman with a doughy complexion and coarse, unkempt hair. Her blood pressure was 88/66 with a pulse of 68 beats/min; respiration was 12 per minute. Chest examination was normal; the abdomen was distended and soft with diffuse discomfort but no local tenderness. Auscultation revealed only scant suggestions of bowel sounds. The examiner detected a colon that was apparently impacted with fecal masses and ordered a barium enema. After considerable difficulty in evacuating a substantial mass of impacted fecal material, a reasonably good filling of the colon was obtained with barium. The x-ray demonstrated a widely distended colon with minimal haustra, diagnosed as megacolon.

The laboratory reported an admission blood analysis of Na, 121 mM; K, 4 mM; Cl, 88 mM; HCO_3, 28 mM; and glucose, 72 mg/100 ml. On rechecking the patient, the physician noted that she was wrapped snugly in her blanket even though the room was comfortably warm. Her chart showed an absence of any fever. He, therefore, ordered laboratory tests for blood cholesterol level and protein-bound iodine. The tests showed a cholesterol of 415 mg/100 ml and a PBI of 2.8 μg/100 ml.

The patient was discharged on appropriate therapy. In follow-up visits over the next 12 months, she reported that she was feeling better than she had in years.

Questions

What specific therapy is indicated for this patient?

Explain the circumstances that brought her to the hospital.

What is the basis for her abnormal blood electrolytes?

Analysis of Case 39

Loss of appetite, lethargy, obesity, constipation, a slow pulse, coarse hair, and a doughy complexion are all characteristic of the myxedematous state, due to hypothyroidism. Of particular physiological interest in this patient is the fact that a secondary complication of her disease brought her to the hospital, where another secondary complication was discovered by the laboratory. The intestinal musculature is among the many functional systems in the body whose activity level is conditioned by the secretion of the thyroid hormone. Hyperthyroidism is characterized by hypermotility of the gastrointestinal tract and a tendency toward diarrhea, while hypothyroidism leads to decreased motility and constipation. It required the discomfort of fecal impaction to prompt this patient to seek medical help. The state of megacolon diagnosed on x-ray was the functional consequence of chronic hypomotility.

The conspicuous laboratory finding in this patient was her severe hyponatremia, which must also have been associated with significant hypo-osmolality. There are two routes for sodium loss from the plasma in hypothyroidism. The myxedema itself, representing the deposition of large amounts of mucopolysaccharides in the skin, takes up sodium in the form of increased tissue fluid as well as from sodium directly bound to the chondroitin sulfuric acid in the mucopolysaccharides. That loss of sodium should be compensated by renal conservation of sodium, were it not for the fact that the second route for sodium loss is the kidneys. Normal activity levels of the tubular sodium pumps and, more specifically, the active transport mechanism stimulated by aldosterone are dependent on adequate titers of thyroid hormone. With thyroid deficiency, the aldosterone sodium conservation mechanism is impaired and sodium wastage occurs.

The renal loss of sodium, which carries significant water osmotically, depletes the extracellular fluid volume, including the blood volume. This lowered blood volume, together with a decreased response to noradrenergic vasoconstrictor mechanisms, would explain the hypotensive blood pressure observed in this patient. A fall in blood volume is sensed by pressure receptors on both the arterial and the venous side of the circulation and signals the hypothalamus to conserve both salt, via the renin—aldosterone system, and water, via the antidiuretic hormone. As pointed out previously, however, the renal response to aldosterone is depressed. The renal response to antidiuretic hormone is apparently well maintained in hypothyroidism. Furthermore, a significant volume depletion takes precedence over osmolality regulation in the control the hypothalamus exerts on antidiuretic hormone release. The result of this precedence is that water conservation persists to sustain the blood volume in spite of the osmotic dilution of body fluids that it produces. As evidence of this derangement in salt and water control, hypothyroid patients in a state of hyponatremia will rapidly excrete a load of salt but will tend to retain an excess load of water.

In spite of the wide spectrum of functional deficiencies that are associated with hypothyroidism, the onset of the disease is so slow and the manifestations are so subtle that patients often do not recognize the extent of their illness until they reach a fairly advanced stage of the disease. This failure to recognize the magnitude of her problem was evident in this patient and accounts for the dramatic improvement in her sense of well-being with the return to a euthyroid state.

40

A 14-year-old boy was referred to the hospital for study because he was underweight, rather listless, and complained that activities of other boys his age were too strenuous for him. His blood pressure was 96/62, his pulse was 80 beats/min, and respiration was 15 per minute. He was observed contentedly watching TV the night of his admission.

The next morning a nurse had difficulty in arousing him and could not readily determine any pulse. A physician was called, who measured a blood pressure of 60/? and detected a weak, thready pulse and distant heart sounds with a rate of about 140 beats/min. Peripheral veins were collapsed. The physician immediately ordered intravenous saline and gave the patient an injection of desoxycorticosterone acetate. In setting up the intravenous, a blood sample was obtained and sent to the laboratory. The patient responded rapidly to the therapy and by noon he was feeling well again. The laboratory reported a hematocrit of 59%, Na of 112 mM, K of 6.8 mM, and glucose of 56 mg/100 ml.

As the physician was leaving the patient's room following a checkup that afternoon, he encountered the patient's mother coming to visit her son. She was carrying a large bag of salted peanuts. In response to the quizzical look of the doctor, she said, "Oh, my boy loves anything salty!"

Questions

What is this patient's problem?

What might have precipitated this acute episode?

Where is the primary lesion?

Analysis of Case 40

The physician was confronted with a youthful patient in a profound state of hypovolemic shock with the knowledge that the patient had previously shown signs of some chronic but not acute illness. He reasoned that there was no basis for suspecting massive hemorrhage and made the correct guess that he was dealing with an Addisonian crisis. The primary lesion was undoubtedly in the adrenal glands because of the deep state of shock. Adrenal cortical deficiencies that are secondary to inadequate ACTH will be dominantly manifest by glucocorticoid deficiencies, since mineralocorticoid secretion will be reasonably well maintained by the renin—angiotensin control mechanism. Also, the high potassium level in the plasma would have been a potent direct stimulus for mineralocorticoid release if the adrenal gland had been functionally responsive. The blood chemistry, showing the low sodium and high potassium, clearly implies a severe mineralocorticoid deficiency in this case.

It is interesting that the patient's history prior to this hospital admission could retrospectively be attributed to a deficiency of glucocorticoids alone. This deficiency would account for the failure to gain weight, the listlessness, and the lack of interest in strenuous activities, all of which relate to the metabolic component of adrenocortical function. This is explained by the fact that deficiencies in aldosterone, the chief mineralocorticoid, can be largely counteracted by adequate dietary intake of salt. With adequate sodium intake to compensate for renal losses, other regulatory systems can maintain salt and water homeostasis.

The admission to a hospital of a patient who was not in a critical condition would not generally be thought of as "stress" in the sense used in adrenocortical physiology. Yet for a boy of this age, the experience may well have had marked effects on his secretion of catecholamines, which are closely coupled to the stress mechanisms associated with adrenocortical responses. A further factor that was probably of major importance in precipitating the Addisonian crisis was that hospitalization deprived this boy of his supply of salt. Further questioning of his mother revealed that, besides his appetite for salty foods, the boy would frequently enter the kitchen at home and eat a handful of salt, poured directly from the salt container. The mother did not recognize this as abnormal behavior in a growing boy. This high level of salt ingestion had been crucially important in protecting the patient from an Addisonian crisis previously.

Abnormal skin pigmentation would have been anticipated in this patient because a deficiency in the response of the adrenal cortex removes inhibitory feedback control on the anterior pituitary gland. The resulting high titer of both ACTH and melanocyte stimulating hormone would jointly stimulate the melanocytes, since both hormones contain the polypeptide chain that is responsible for melanocyte stimulation. For unknown reasons, some patients fail to exhibit this melanocyte response in spite of high titers of ACTH.

41

A 58-year-old male was admitted to the hospital complaining of shortness of breath, fatigue, and muscular weakness. He reported that two
years ago he had been told that he had high blood pressure following
a routine examination.

This rather obese and obviously apprehensive patient had a blood
pressure of 210/90 and an irregular pulse of about 90 beats/min. He
had dyspneic respirations at a rate of 38 per minute. His heart was enlarged and there was dullness at both lung bases. Neck veins were distended and the liver was significantly enlarged. He had marked ankle
edema. A chest x-ray confirmed the enlarged heart and showed pulmonary congestion. An electrocardiogram demonstrated atrial fibrillation
and left-axis deviation.

The patient was digitalized and placed on bed rest. For three
days his condition remained essentially unchanged. Though he ate well
and was reasonably comfortable, atrial fibrillation persisted, he required
three pillows at night to avoid excessive dyspnea, and he failed to show
any diuresis. On the fourth day the nurse who bathed him reported
that the patient was too weak to sit up and have his back washed.

Clinical laboratory data were as follows:

Na	144 mM	Cl	110 mM
K	3.4 mM	HCO_3	21 mM
H^+	35 nM	Protein-	8.7 μg/100 ml
Glucose	107 mg/100 ml	bound iodine	
Cholesterol	128 mg/100 ml	24-hour urine:	

      4.3 gm of creatinine

      13.6 mg of 17-hydroxycor-
      ticoids (normal is 9)

The patient was referred to the Endocrinology Department for
further study and radioiodine treatment.

Questions

What underlying disease can you identify in this patient?

How does this relate to his congestive heart failure?

What accounts for his muscular weakness?

What accounts for his obesity?

Analysis of Case 41

Assuming no unusual iodine intake, this patient's high protein-bound iodine and low cholesterol identify his basic problem as one of hyper-thyroidism that first became evident when it precipitated this bout of congestive heart failure. His hypertension was chiefly systolic, with only a marginal elevation of the diastolic pressure in spite of the rapid pulse. These data indicate a large stroke volume and a high cardiac out-put, suggesting a hypermetabolic state. Thyroid hormone has a power-ful action to augment oxidative metabolism throughout the body; this metabolic activity releases local dilator metabolites to open up the peripheral circulation and produce a high blood flow and a large venous return. The metabolic effect to stimulate blood flow acts synergistically with thyroxin potentiation of sympathetic stimulation of the heart, maintaining a high cardiac output.

Unfortunately, this high output state exerts a toll on myocardial function. As with thyroid effects on other body tissues, the increased energy production by thyroxin stimulation is not coupled with an equivalent increase in useful work output, and myocardial efficiency falls. The heart, which is sensitized to adrenergic stimulation by thy-roid hormone, becomes very susceptible to the development of ectopic foci of stimulation and arrhythmias, as demonstrated by the atrial fibrillation shown by this patient. By disrupting the normal sinus rhy-thm and uniform filling and emptying cycles, atrial fibrillation further impairs the mechanical efficiency of the heart. Complicating the treat-ment of this condition is the fact that both the antiarrhythmic proper-ties and the inotropic action of the cardiac glycosides are ineffective in the presence of excess thyroxin. This is probably explained by the fact that these cardiac drugs exert their primary action by increasing the entrance of Ca^{++} ion into the myocardial cell, an effect that has already been achieved by the combined action of thyroxin and catechol-amines. This combination of circumstances can lead to a form of con-gestive heart failure that is refractory to treatment except for specific steps to remove the excess thyroxin. Since this refractoriness to therapy also makes these patients poor surgical risks, most physicians would elect radiation techniques to suppress the thyroid hormone secretion.

While the muscular weakness observed in hyperthyroidism is not fully understood, two important factors can be identified. (1) Excess thyroid hormone acts to block the phosphorylation of muscle creatine, the major energy store of skeletal muscle; and both creatine and other muscle constituents are lost from the muscle, and increased excretion of urinary creatinine occurs. (2) Extracellular potassium tends to be low in hyperthyroidism. This factor could relate to increased cellular uptake of K^+ because of heightened cellular metabolic activity, and it could also reflect renal losses that are secondary to the increased secretion of both glucocorticoids and mineralocorticoids in the hyperthyroid state. Lowered extracellular K^+ depresses the excitability of peripheral tissues even though the hyperthyroid state is usually characterized by hyperexcitability of the central nervous system. This patient showed evidence of these chemical disturbances, though not of sufficient magnitude to explain readily the degree of weakness observed.

This patient was obese for the same reason that any other patient develops obesity: intake of food in excess of metabolic requirements. Note that this patient was observed to be maintaining a good appetite in the hospital under circumstances in which other patients might have little interest in food. Many patients with hyperthyroidism do not increase their food intake sufficiently to keep up with their hypermetabolic state and, hence, will tend to lose weight. In other patients, their appetite drive keeps ahead of their hypermetabolism and they can gain weight in spite of their increased calorie wastage.

42

A 48-year-old typesetter was admitted to the hospital with anuria, reporting that for two days he had been vomiting and passing black stools. He had a long history of dyspepsia with a known bleeding peptic ulcer for seven years. Surgery for the ulcer had been advised repeatedly, but the patient had refused it and elected self-medication, relying on the ingestion of large amounts of milk and proprietary alkali products to control his symptoms.

Physical examination disclosed an apprehensive, thin, pale individual complaining of weakness, thirst, and epigastric pain. His blood pressure was 90/60, and his pulse was 130 beats/min. Chest examination was normal. The laboratory data were as follows:

Na	152 mM	Cl	81 mM
K	2.8 mM	HCO_3	48 mM
H^+	25 nM	PO_4	1.9 mg/100 ml
Ca	2.7 mM	Hematocrit	21%
BUN	61 mg/100 ml	Hemoglobin	7.4 gm/100 ml

He was given intravenous fluids in the form of 5% glucose, plasma, and whole blood, and his condition progressively improved. His urine flow was restored on his third hospital day. On his sixth hospital day urine output had increased to 1450 ml with a specific gravity of 1.010. Creatinine clearance was 82 ml/min. The urine had a cloudy appearance that cleared on acidification, and only traces of albumin were present. Gastric surgery was again advised. He refused and left the hospital against medical advice.

He was readmitted to the hospital the following year with the complaint of cough and chest pain. His blood pressure was 115/75, and his temperature was 39.2°C. A diagnosis of right lower lobe pneumonitis was made and he was treated with antibiotics. Under this therapy his fever rapidly subsided and within five days his lungs had cleared. Laboratory data at this time were as follows:

Na	145 mM	Cl	92 mM
K	4.8 mM	HCO_3	34 mM
H^+	30 nM	PO_4	1.6 mg/100 ml
Ca	2.9 mM	Hematocrit	32%
BUN	96 mg/100 ml	Hemoglobin	9.8 gm/100 ml
Creatinine	7.3 mg/100 ml		

His urine volumes were 100 and 150 ml for his first two hospital days. By the time of his discharge 12 days after admission, his BUN had dropped to 33, creatinine to 4.9, and his creatinine clearance was 64 ml/min with a urine specific gravity of 1.011.

Three years after his original admission he was again brought to the hospital with gastric bleeding. On this occasion he was clearly in a state of shock with a blood pressure of 72/42. He was treated initially with transfusions and intravenous fluids. Admission laboratory data were as follows:

Na	135 mM	Cl	84 mM
K	3.0 mM	HCO_3	37 mM
H^+	30 nM	PO_4	1.2 gm/100 ml
Ca	3.3 mM	Hematocrit	18%
BUN	122 mg/100 ml	Hemoglobin	6.4 gm/100 ml

He was anuric for three days and his BUN rose to 173 mg%. Urine flow was then restored with a progressive fall in his BUN.

On this admission he was convinced of the urgency for surgical treatment of his ulcer. Because of the precarious state of his health, surgery was restricted to a gastroenterostomy and a vagotomy. Initially, the patient responded well to the surgery and maintained urine outputs between 1500 to 3000 ml/day. A laboratory test on his fifth postoperative day showed a plasma Na of 122 mM, which prompted the administration of 1 liter of 2% NaCl. Twenty-four hours later, plasma Na was 125 mM.

This latter finding and the improved condition of the patient prompted a more thorough study of the patient's renal function. His urine showed a few blood cells and occasional casts, an H^+ concentration of 96 nM, a specific gravity of 1.010, traces of albumin, and no glucose. A phenol red test demonstrated excretion of 4% of the injected dye in the first hour (normal is 50%). Withholding of fluids for a 24-hour period elevated the specific gravity of the urine from 1.010 to 1.012. On another day a 24-hour urine collection showed a volume of 1840 ml, containing Na in a concentration of 117 mM; creatinine, 62 mg%; and urea, 830 mg%. Blood chemistries on that day were as follows:

Na	124 mM	Cl	87 mM
K	5.2 mM	HCO_3	21 mM
H^+	42 nM	BUN	48 mg/100 ml
Ca	3.2 mM	Creatinine	3.4 mg/100 ml
PO_4	1.1 mg/100 ml		

On the basis of these findings and a careful review of the history, x-ray examination of the long bones was ordered. The x-rays revealed significant patchy areas of demineralization and rarefaction of the bone substance. On this basis a further diagnosis was made that was confirmed during an exploratory operation. The patient was discharged with strict dietary instructions, including vitamin D supplements.

At a follow-up examination three years later, the patient was in reasonably good health with no recurrence of his ulcer problem. A BUN at this time was 26 mg/100 ml.

Questions

In untangling the interrelated factors in a case as complicated as this, it is important to dissect the different facets of the problem. The basic physiology of peptic ulcers has been discussed previously (cf. Case 19). Why would ingestion of milk and alkali relieve ulcer symptoms?

What factors could have contributed to the elevated BUN observed at the time of the first hospital admission?

Review the total history from the standpoint of renal function; what evidence of impaired renal function can you identify?

What consistent abnormality in blood chemistry would have prompted the x-ray examination of the skeleton?

How might the chronic ingestion of large amounts of milk and alkali relate to this finding?

What physiological control system has been severely taxed throughout the course of this illness?

What type of pathology was encountered in the last surgical exploration, and how would this relate to his ulcer problem?

Can one be confident that vitamin D will be effective therapy in a patient with severe renal disease?

Analysis of Case 42

This case history describes a problem whose complexity evolved from the interrelationships between body functions, whereby one functional derangement cascaded into a series of secondary functional disturbances. At the outset, the history of this patient presents the problem of peptic ulceration that is refractory to conservative treatment. Though it is not possible to identify the underlying cause of ulcer formation in many patients, an important group of those ulcer patients who are refractory to conservative treatment have an abnormally high secretion of gastrin, which is the major hormonal regulator of gastric acid secretion. A basic question to consider, therefore, is whether this patient developed some source for increased gastrin stimulation.

An additional problem is clearly evident in the blood chemistries recorded at the time of his first hospital admission, which showed a high BUN. Several factors would have elevated this patient's BUN. During an acute gastric upset, intake of both food and fluid is sharply curtailed. To satisfy body energy requirements without caloric intake, glucose equivalents are mobilized from protein, resulting in an overload of urea production to dispose of the amino nitrogen. Excess amino nitrogen would also be entering the urea cycle from digestion of the blood that had passed into the intestinal tract. In addition, the lack of fluid intake and the excess fluid losses due to the vomiting would act to concentrate body fluids and thereby exaggerate the rise in urea concentration. Finally, renal function is maintained in an optimal homeostatic range by autoregulatory control as long as arterial blood pressure is reasonably well maintained. The hypotension this patient exhibited by the time he entered the hospital, due to blood and fluid loss, led to vasoconstrictive throttling of renal blood flow and the consequent state of anuria. Elimination of urea excretion would further elevate the BUN. Yet in spite of these several factors acting to elevate urea levels, it must be recognized that urea is a small molecule whose solubility properties make it permeable to almost all membranes, and hence its dilution volume encompasses the entire reservoir of body fluids. Such a rise in the urea concentration in this large reservoir cannot reasonably be attributed entirely to these aspects of the patient's acute illness and argues for the existence of underlying chronic renal disease. Confirmation of this was found in the creatinine clearance, attesting to a somewhat depressed glomerular filtration rate.

A similar situation was observed when the patient was admitted to the hospital for his acute pulmonary disease. A febrile illness accelerates protein catabolism and has a tendency to elevate the BUN, and the relative state of oliguria would augment this rise. However, the observed BUN was far in excess of what could be accounted for on this basis. At this time the creatinine clearance was somewhat more significantly depressed. The uremia observed at the time of the third hospital admission, though accentuated by the same prerenal factors discussed in reference to his first admission, removes any doubt as to the severity of the renal component of his disease. Until the difficulty was encountered, postoperatively, of maintaining his sodium balance, however, other problems demanded priority over concern for his renal disease.

The surgical treatment of the patient was a relatively minimal attack on his ulcer problem. The vagotomy to reduce gastric acid secretion and the gastroenterostomy to short-circuit most of the acid chyme away from the lesion would, it was hoped, allow sufficient healing to curb the immediate risk of a fatal hemorrhage.

The postoperative renal study demonstrated gross deficiencies in both glomerular filtration and tubular secretory function. Also, the inability to concentrate the urine suggests an almost total functional loss of the medullary countercurrent multiplier of the loops of Henle. This would account for the last set of data given, which translate into a urea clearance of 22.1 ml/min, almost equaling the creatinine clearance of 23.4 ml/min. Normally, urea clearance does not approach the creatinine clearance so closely because of back diffusion of both water and urea into the medullary pyramids in the process of concentrating the urine. Failure of the medullary sodium pumps and the relatively high flow rate through the residual functioning cortical nephrons would explain the salt-wasting phenomenon. The most powerful salt pumps are those found in the medullary nephrons. The action of the sodium pumps in the cortical nephrons becomes compromised when they are presented with a large load of filtrate because the tubule cells reabsorb fluid more rapidly than the peritubular capillaries can transport this fluid away. Fluid, therefore, accumulates between the cells of the tubular epithelium and tends to leak back into the tubular lumen through the "tight junctions" between cells. The deficiency in the tubular pumping mechanism for sodium in this man's kidneys is especially evident in the data demonstrating a urinary concentration of sodium, which is almost the same as the plasma concentration of sodium in spite of the state of hyponatremia that existed.

A review of the urinary findings shows that, throughout this patient's history, urines had very consistently run a specific gravity of approximately 1.010, corresponding with isotonic urine and demonstrating a failure of the medullary concentrating mechanism. The clue to the etiology of this deficiency was seen at the time of the first hospital admission in the report of a cloudy urine that cleared on acidification. This is an indication that the urine contained colloidal calcium salts. Correlating with this was a very consistent tendency for plasma calcium levels to be somewhat high and for phosphate to be low. This would be explained by a deficiency in renal reabsorption of phosphate which, in the presence of increased filtered calcium, would lead to precipitation of calcium salts in the tubules as the urine became concentrated. During earlier phases of this patient's disease, this concentration would have become greatest in the medullary loops of Henle at the time that they were still functional, and the resultant precipitation would have produced mechanical blockage and destruction of tubular elements. Though there was no history of gross kidney stones, microscopic urinary calcinosis would also account for the more generalized destruction of renal tubular elements as the disease progressed.

Evidence of persistent, low plasma phosphate, high plasma calcium, and demineralization of the skeleton suggests hyperparathyroidism. This was confirmed by the surgical removal of hypertrophic parathyroid glands. It is a matter of speculation as to just how this vicious cycle is initiated. The parathyroid glands are stimulated to mobilize bone calcium when plasma calcium levels fall. While milk is commonly regarded as a food that supplies an abundance of calcium, the simultaneous ingestion of alkali renders the calcium insoluble and prevents its absorption. Moreover, the high phosphate content of milk fosters the precipitation of calcium in an alkaline medium and causes a net movement of calcium out of the blood and into the intestine. This fall in blood calcium following the ingestion of milk and alkali stimulates the parathyroid glands to release parathormone to mobilize calcium from the bone stores. The vicious cycle is completed by the fact that the resulting rise in blood calcium, due to the action of parathormone, stimulates the release of gastrin. In normal homeostatic regulation of calcium, this coupled hormone response increases gastric acidification and facilitates calcium absorption from the intestine. This synergistic action is thwarted by the simultaneous ingestion of alkali. Furthermore, since the delayed hormone response would not act until well after the alkali left the stomach, the gastrin stimulation causes gastric hyperacidity. This condition would aggravate the peptic ulceration and encourage the patient to ingest more milk—alkali, leading to more parathyroid stimulation and more gastric hyperacidity.

Still further complications arise when this disturbed calcium metabolism induces renal damage, since the kidneys appear to be the chief organ for metabolizing calcitonin, the hormonal antagonist to parathormone. Thus there is a total breakdown of the hormonal mechanisms for homeostatic regulation of calcium. In whatever fashion this vicious cycle may have been initiated, the end result is hypertrophy of the parathyroid glands and a chronic state of hyperparathyroidism.

It may seem rather remarkable that this complex derangement could be significantly improved by the surgical substitution of hypoparathyroidism. Loss of the parathormone, however, eliminated the stimulus for gastrin release and, together with the vagotomy and gastroenterostomy, apparently permitted healing of the patient's ulcer. In reference to the patient's renal function, it should be appreciated that the period of shock at the time of his third hospital admission would have superimposed some acute ischemic tubular damage on his chronic renal disease and thereby exaggerated the extent of his renal insufficiency. During the convalescent period, there is substantial recovery of the renal dysfunction that is attributable to this acute ischemic injury. Indeed, restoration of a reasonable level of kidney function was essential for the success of the vitamin D therapy that was used to compensate for the loss of parathormone. The kidneys perform an essential hydroxylation of vitamin D to convert it into its biologically active form. Failure of this mechanism can impose a still further disruption in calcium metabolism in patients with severe renal disease.

43

A 34-year-old housewife, while attempting to cross a busy intersection, barely missed being struck by a taxi and collapsed on the street. She was rushed to a nearby hospital in an ambulance. By the time she reached the hospital, she had regained consciousness. Examination in the emergency room revealed nothing of significance. (In a later history, she confessed that she had a severe headache at the time that had persisted for several days, but she had concealed this fact for fear of being detained at the hospital.) For the following four months she noted an absence of menstrual periods and sought help from her physician. After a routine examination, he placed her on a program of diethylstilbestrol therapy and thyroxin. The patient discontinued the thyroxin voluntarily because it gave her annoying palpitations.

Six months later she was admitted to the hospital with vomiting and abdominal pain. She reported that her entire family had been ill with an apparent case of food poisoning for the previous 24 hours. She appeared to be severely depressed. Her pulse was 105 beats/min, her blood pressure was 98/64, and respiration was 33 per minute. Blood chemistries reported by the laboratory were as follows:

Na	122 mM	Cl	82 mM
K	5.9 mM	HCO_3	29 mM
H^+	35 nM	P_{CO_2}	43 mm Hg
Glucose	78 mg/100 ml	Cholesterol	401 mg/100 ml
BUN	27 mg/100 ml		

She was given appropriate treatment for her acute illness and a thorough workup to attempt to diagnose the etiology of her underlying disease. [131]I uptake by the thyroid was 8% of a tracer dose in 24 hours (normal is 25%). She showed no evidence of pigmentation or exophthalmus, and an x-ray of the skull demonstrated a normal sella turcica. She was discharged with a combination of cortisone and thyroxin therapy.

She was readmitted to the hospital eight months later complaining of headache and fever. Her physical appearance at this time showed a striking change. Her face had filled out remarkably, with puffy eyes and a double chin. She had adopted a somewhat stooped posture that was exaggerated by an accumulation of adipose tissue that had developed between her shoulders. In contrast, her extremities were gaunt and she complained of weakness and easy fatigability. She had a hot, dry skin with an oral temperature of 39.6°C. Her blood pressure was 220/90 with a pulse of 110 beats/min. Laboratory tests gave the data shown as follows:

Na	148 mM	Cl	117 mM
K	3.2 mM	HCO_3	26 mM
H^+	36 nM	P_{CO_2}	37 mm Hg
Glucose	185 mg/100 ml		

A 24-hour urine was found to contain 1 mg of 17-ketosteroids (normal is 10) and 32 mg of urinary corticoids (normal is 12).

The therapy that she had been taking was immediately discontinued and an intensive investigation was undertaken to identify a possible source of infection. Her white blood cell count was normal except for a low eosinophil count. With a failure to identify any specific focus of infection, she was placed on a broad-spectrum antibiotic.

On her fourth hospital day her 17-ketosteroids were again measured and found to be 3 mg/24 hr. Administration of ACTH on her fifth hospital day raised her 17-ketosteroids to 15 mg/24 hr. Her fever continued to rise in spite of her therapy and was 40.8°C on her fifth hospital day. At this time her blood pressure was 135/80 with a pulse of 124 beats/min.

On her sixth hospital day her fever rose to 41.6°C and intensive efforts were made to lower her temperature with alcohol baths and cold-water enemas. In spite of this, her temperature reached 44.5°C on the morning of her seventh hospital day. At this time her pulse rate was 142 beats/min and her blood pressure was 70 systolic; the diastolic not identifiable. She died in the latter part of the morning.

An autopsy was obtained.

Questions

What was the diagnostic problem at the time of this patient's initial complaint of amenorrhea?

What features of this patient's response to the food poisoning episode were not typical?

What accounted for the abnormal steroid excretion at the time of her second hospital admission?

What primary site for her disease is suggested by the terminal febrile course?

Analysis of Case 43

This patient portrayed the developing picture of panhypopituitarism. The female reproductive cycle tends to be the first manifestation of such a problem, because minor disturbances in relative hormone titers can lead to gross disturbances in function. The very sensitivity of this cycle to a wide assortment of influences, however, makes it difficult to attempt a definitive diagnosis when amenorrhea is the only recognized problem. There was no evidence, for example, that the physician who originally examined her had any specific justification for prescribing thyroxin, but suboptimal thyroid function is one of the many factors that can upset the precise tuning of the menstrual cycle.

At the time of her first hospital admission, the patient was clearly overreacting to a stress that she had shared with other members of the family. A mild metabolic alkalosis following a bout of vomiting was to be expected, but the hyponatremia, hyperkalemia, and hypotension suggested inadequate adrenocortical response to the stress. The high cholesterol level could have been related to a high titer of ACTH responding to the absence of appropriate feedback inhibition from the adrenal cortex, but the reduced iodine uptake by the thyroid indicates that the high cholesterol was more likely to be a manifestation of a deficiency of thyroid hormone, whose concentration influences the rate of hepatic uptake of cholesterol. The evidence, therefore, defines deficiencies in three of the endocrine organs under pituitary control: the thyroid, the adrenal cortex, and the ovarian—uterine system. At the same time, there is lack of evidence of skin pigmentation due to the melanocyte-stimulating property of ACTH if the adrenal deficiency were in the peripheral gland. Therapy was directed toward correcting the two hormone deficiencies that were essential to maintenance of health: the hormones of the adrenal cortex and of the thyroid.

The second hospital admission presented two new aspects: (1) the combination of headache and fever without any obvious cause, and (2) evidence of chronic overdosage with cortisone, producing a Cushing-like syndrome. The latter was manifested in the classic Cushing physique, together with hypertension, hyperglycemia, and eosinopenia. The cortisone therapy caused the excessive excretion of urinary corticoids, while the endogenously produced steroids were represented in the very low 17-ketosteroid fraction. The test administration of ACTH evoked an impressive rise in the 17-ketosteroid excretion, confirming the assumption that there was no primary deficiency in the adrenal cortex.

The pattern of the progressive fever, apparently unrelated to infection, assumed the characteristics of a pathological hyperthermia, and shifted the focus of attention from the pituitary to the hypothalamus. The patient now exhibited a deficiency of heat-loss functions, which could correlate with a deficiency of the hypothalamic releasing factors that play a key role in the control of gonadotrophin, thyrotrophin, and adrenocorticotrophin secretion. The panhypopituitarism would have been secondary to this primary hypothalamic deficiency. It should be noted that at no time in her history had this patient's endocrine deficiencies been extremely severe. This relates to the fact that, although the hypothalamus is a major controller of the pituitary, there is also some direct feedback control on the pituitary itself. At the time of the patient's admission with the Cushing-like manifestations, for example, the extremely low level of the endogenous 17-ketosteroid excretion was a reflection of the feedback inhibition of the high levels of exogenous cortisone, a feedback circuit that was probably acting directly on the pituitary.

At autopsy, the only significant finding was evidence of both old and recent hemorrhage into the substance of the hypothalamus from a small cerebral aneurysm. The pituitary showed some atrophic changes but no specific pathology. Both clinically and pathologically, the antidiuretic function of the posterior pituitary had been preserved. It may be conjectured that an initial hemorrhage from the aneurysm occurred at the time that the patient was frightened by the taxi. A contributing factor to the subsequent fatal hemorrhage was undoubtedly the hypertension produced by the excessive cortisone therapy.

44

A 24-year-old married waitress was brought to the hospital complaining of severe abdominal pain. She reported that she had felt well until 3 hours previously, when a sharp pain had developed in her side.

She was a young woman in acute distress, extremely pale, with cold, moist skin. Her blood pressure was 60/40 with a thready pulse of 128 beats/min. Her respiration was 24 per minute; temperature was 38.1°C. Her hematocrit was 29%. Examination of her chest was normal. Abdominal examination revealed a rigid abdomen that was extremely tender throughout, with the pain perhaps slightly greater in the lower two quadrants. A few bowel sounds were heard. Pelvic examination gave evidence of menstrual flow and the impression of a normal uterus, although the patient was too tender to conduct a satisfactory examination.

The patient stated that she had experienced menarche at the age of 13 and that her subsequent menstrual history had been somewhat irregular, with periods at intervals of 30 to 35 days and sometimes longer. She had not used contraceptive medication for several years because of annoying side effects; she and her husband had relied on condoms for birth control. Her present menses had started that morning, following a period of seven weeks from her previous menstruation. She had noted some detectable breast enlargement during the previous few weeks, which had been typical whenever her period had been significantly delayed. A radioimmunoassay test for pregnancy was found to be negative.

A blood sample was sent to the laboratory for crossmatching and the patient was taken directly to the operating room for an exploratory laparotomy.

Questions

Would you deem a patient with a blood pressure of 60/40 to be a good surgical risk without preoperative treatment for her shock?

What effect does intense visceral pain have on blood pressure? Could that explain the blood pressure findings in this patient?

Of what significance was the hematocrit reading?

How would you explain the breast enlargement reported by this patient?

Of what significance was the negative pregnancy test?

Analysis of Case 44

This young woman entered the hospital in a severe state of shock with an acute abdominal problem; she clearly represented a surgical emergency (cf. Case 29). However, the physician must carefully weigh the urgency of immediate surgery against the prophylactic value of treating the shock and attempting to gain a more substantial basis for a diagnosis before surgical exploration. In addition to clues obtainable from the patient's history, analysis of the physiological data will aid in making this decision.

Most patients who present with an acute abdominal problem will exhibit some hypotension and other signs of shock, because an inflammatory process sufficiently severe to produce peritoneal signs will reflexly inhibit vasoconstrictor tone and cause a generalized vasodilation. In this state, which has been aptly described as *primary shock*, blood flow and tissue perfusion will generally be high. Primary shock, therefore, does not represent the life-threatening emergency of hypovolemic shock and generalized tissue ischemia. With moderate degrees of primary shock, the clue to the patient's condition is found in the blood pressures, which exhibit reasonably good pulse pressures in spite of the low absolute pressure level. A patient with a substantial pulse pressure must have a good cardiac output and, therefore, is maintaining tissue perfusion. Typically such patients will have a warm skin as direct evidence of this vasodilated state. Unfortunately these criteria are not of value in more severe states of primary shock because the vasodilated state includes sufficient dilation of venous reservoirs to compromise venous return to the heart, and such profound dilation of resistance vessels that there is an inadequate arterial pressure head to maintain good perfusion of the skin in spite of its dilated state.

In assessing severe hypotensive states, the hematocrit provides a valuable clue. The reflex dilation of the arterioles in primary shock increases capillary hydrostatic pressure, which favors the transudation of fluid from the capillaries into the interstitial space. This loss of fluid from the plasma compartment elevates the hematocrit. In hypovolemic shock, by contrast, the hematocrit falls. This fall is a result of the reflex vasoconstriction induced by the fall in blood pressure and the consequent reduction of the pressure stimulus on the arterial baroreceptors, by which feedback inhibition is removed from the sympathetic constrictor system. The consequent constriction of the arterioles lowers the capillary hydrostatic pressure, shifts the capillary equilibrium to favor the oncotic pressure of the plasma proteins, and causes resorption of fluid from the interstitium into the plasma. This autotransfusion of fluid provides some compensation for loss of circulating blood volume and is manifest clinically by a fall in the hematocrit.

This patient, with a cold, clammy skin and a low hematocrit, gives clear evidence of being in a vasoconstricted state with severe hypovolemia. The slight temperature elevation could be a manifestation of cutaneous vasoconstriction interfering with heat loss. The picture, therefore, suggests massive internal hemorrhage, in which case transfusions to elevate the blood pressure could serve to increase the bleeding and further compromise, rather than benefit, the condition of the patient.

There were minimal clues to identify the cause of this patient's shock, other than knowledge of statistical probabilities of what intra-abdominal catastrophes might be anticipated in a young woman. The breast changes described by the patient related to the action of estrogens, which stimulate the proliferation of the duct system, and to progesterone, which stimulates the development of the alveolar structure. The waxing and waning of these hormones, as a result of the secretory activities of the follicle and the corpus luteum, produces some minimal changes in the breast during normal menstrual cycles, and more noticeable changes during protracted menstrual periods. The functional prolongation of the luteal phase of the menstrual cycle in some nonpregnant women is not precisely understood, but it may relate to a disturbance in a negative feedback loop from the uterine endometrium which normally accelerates the regression of the corpus luteum.

Alternatively, the prolonged menstrual interval in this patient could have been caused by pregnancy and maintenance of a functional corpus luteum by secretions of the chorionic membranes of the developing placenta. The presence of this chorionic gonadotrophin is the basis of tests for pregnancy. The negative pregnancy test in this patient was of little significance, since she was at an early stage of pregnancy where sufficient chorion development to yield a positive test might not have been achieved. Furthermore, the diagnosis being considered was an ectopic pregnancy that had failed to implant in the uterus. The development of placental structures would, therefore, have been compromised.

The laparotomy disclosed massive hemorrhage into the abdominal cavity that was traced to bleeding from the rupture of a highly inflamed fallopian tube. The fertilized ovum had failed to move into the uterus, and tubal implantation had occurred. The fallopian tube could not mobilize the structural support to sustain the pregnancy, and rupture and hemorrhage resulted. Obviously, the vaginal bleeding noted by the patient and observed at the time of the pelvic examination was not a true menstrual flow.

Appendixes

Conversion of Hydrogen Ion Concentration to pH Scale

H^+nM	pH	H^+nM	pH
10	8.00	56	7.25
12	7.92	58	7.24
14	7.85	60	7.22
16	7.80	62	7.21
18	7.74	64	7.19
20	7.70	66	7.18
22	7.67	68	7.17
24	7.62	70	7.15
26	7.59	72	7.14
28	7.55	74	7.13
30	7.52	76	7.12
32	7.49	78	7.11
34	7.47	80	7.10
36	7.44	82	7.09
38	7.42	84	7.08
40	7.40	86	7.07
42	7.38	88	7.06
44	7.36	90	7.05
46	7.34	92	7.04
48	7.32	94	7.03
50	7.30	96	7.02
52	7.28	98	7.01
54	7.27	100	7.00

Normal Blood Values

Na	140 mM
K	5 mM
H^+	40 nM
Ca	2.5 mM
Cl	102 mM
HCO_3	24 mM
Anion gap $(Na + K) - (Cl + HCO_3)$	20 mM
PO_4, measured as P_i	3.5 mg/100 ml
Glucose	90 mg/100 ml
Urea (BUN)	12 mg/100 ml
Creatinine	2 mg/100 ml
Lactate	8 mg/100 ml
Hemoglobin	15 gm/100 ml
Albumin	4.5 gm/100 ml
Globulin	2.5 gm/100 ml
Bilirubin (total)	0.3 mg/100 ml
Cholesterol	200 mg/100 ml
Protein-bound iodine	6 μg/100 ml
Alkaline phosphatase	3 units

Index